Science for a Fragile World

Science for a Fragile World

ROBERT NORTHCOTT

OXFORD
UNIVERSITY PRESS

OXFORD
UNIVERSITY PRESS

Great Clarendon Street, Oxford, OX2 6DP,
United Kingdom

Oxford University Press is a department of the University of Oxford.
It furthers the University's objective of excellence in research, scholarship,
and education by publishing worldwide. Oxford is a registered trade mark of
Oxford University Press in the UK and in certain other countries.

Published in the United States of America by Oxford University Press
198 Madison Avenue, New York, NY 10016, United States of America.

British Library Cataloguing in Publication Data
Data available

Library of Congress Control Number: 2025935330

ISBN 9780192849083

DOI: 10.1093/9780191944352.001.0001

Printed and bound by
CPI Group (UK) Ltd., Croydon, CR0 4YY

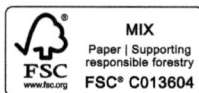

MIX
Paper | Supporting
responsible forestry
FSC
www.fsc.org
FSC® C013604

The manufacturer's authorised representative in the EU for product safety is
Oxford University Press España S.A., Parque Empresarial San Fernando de Henares,
Avenida de Castilla, 2 – 28830 Madrid (www.oup.es/en or product.safety@oup.com).
OUP España S.A. also acts as importer into Spain of products made by the manufacturer.

Contents

Acknowledgments

This book, like most books, is the product not just of its author but also of many other people, to all of whom I am grateful. Adrian Currie and Julian Reiss generously read and gave very helpful feedback on the entire draft manuscript. Others read and gave very helpful feedback on large sections of it, notably Jacob Stegenga and many of his students, Nancy Cartwright, and before that, several anonymous referees for journals and book chapters. All mistakes that remain are my own, of course. More generally, like anyone, my ideas are the product of many influences that I am no doubt unaware of or have forgotten. Among those whose work I know has been especially helpful for me are Nancy Cartwright, Julian Reiss, Petri Ylikoski, and Anna Alexandrova. I benefited from a reading group at Cambridge on Donatella della Porta's "Social Movements, Political Violence, and the State" and, also at Cambridge, from many meetings of the "Calculating People" reading group run by Anna Alexandrova.

The ideas in the book were refined and improved by discussion with many other philosophers and by attentive feedback from audiences at my own presentations. In this way, I have benefited from the wide world of academic events that many of us are lucky to enjoy, on the back of huge efforts by their organizers. These include events at, in roughly chronological order, the following venues (often multiple times): Birkbeck, London School of Economics, Kent, Oxford, Johannesburg (online), Bielefeld, Kings College London, Durham, Aarhus, Washington University St. Louis, St. Louis University, Valencia, Helsinki, Cambridge, and Oslo. In addition, at the following learned society conferences: the British Society for the Philosophy of Science at Cambridge, Oxford, Durham, and Bristol; the Philosophy of Science Association at Chicago; the International Network for Economic Method at San Sebastian and summer school at Como; the American Philosophical Association at San Diego; the Congress on Logic, Methodology and Philosophy of Science and Technology at Prague; the European Philosophy of Science Association at Geneva; and the Society for the Philosophy of Science in Practice at Ghent. It was a question from Donal Khosrowi at a workshop in Bielefeld that first triggered the train of thought that led me to fragility and to that becoming the book's central organizing concept.

As a student, I benefited from some very high-level teachers at the London School of Economics, notably Elliott Sober, Nancy Cartwright, and the late Colin Howson. I am also grateful for much supererogatory support I received while a visiting student for two years at the University of California San Diego, notably from Craig Callender, Jonathan Cohen, and Agustín Rayo. As a lecturer and professor, I have been lucky to have consistently warm and supportive colleagues, first at the University of Missouri–St. Louis, and latterly at Birkbeck, University of London.

Staff at Oxford University Press have worked hard to bring this book to production, especially, at various stages, Peter Momtchiloff, April Peake, Mirza Rizwaan Ahmad, and Lavanya Nithya.

Lastly, I want to thank my family, without whom the book would not have been written nor feel half as worthwhile. I dedicate this book to them. That begins with my parents, Jim and Jacquin Northcott, for lifelong, unfailing love and support that has made everything possible. It extends to Anna Alexandrova: not only is she a wonderful partner and coparent, but I also have the great good fortune of talking every day to a fantastic philosopher of science in my own household. And it extends to my children, Andrei and Michael—with all love from dad, and also with all thanks for putting up with me repeatedly disappearing to work on this book. I hope it is worth it!

1
Introduction

1.1 Two worlds

Imagine a world of order. Causal relations are reliable; mechanisms, structures, and functional dependencies persist; and laws are unchanging. How best to investigate such a world? By uncovering these cogs and wheels of nature, confident that they work widely: chemical salts combine, cells respire, and subatomic particles interact, all in regular ways. Knowledge of such things, accumulated over generations, is the route to remarkable power: technology works by exploiting these cogs and wheels. And they are our key, too, for predicting and explaining things. Science in such a world is an inspiring project, familiar from textbooks and popular image. Beneath the messy imperfection around us is an underlying order, and discovering this order is science's mission.

Now imagine a second world, this time one in which laws and causal relations are fragile, winking in and out like bubbles in a boiling soup. This world is very different. Just because one thing causes another over there, that does not mean it will over here. When predicting and explaining, we have no general guides to lean on and instead are forced knee-deep into idiosyncratic local detail. No trustworthy cogs and wheels, rather each time a new look. Artifacts lose their power because the building blocks on which they depend now hold only unreliably. Progress in this world is still possible and remains vital to our fortunes, but it is frustratingly piecemeal and patchwork.

Which of these worlds is our world? As we will see, our world is a sinuous, interlocking mixture of the two. Fragility and stability cross-cut almost all domains, and how best to navigate a way through is an intricate matter. But for many of the things we care about most, our world is fragile, and we need to reckon with that.

Science for a Fragile World. Robert Northcott, Oxford University Press. © Robert Northcott 2025.
DOI: 10.1093/9780191944352.003.0001

1.2 Car radiators

Begin with a simple, artificial case for clarity. Imagine that an old-style car radiator has cracked because water left in the radiator froze and expanded during a cold night—to borrow Carl Hempel's famous example.[1] The connection between the cold night and the cracking of the radiator is stable, based on the stable facts that ice has lower density than water, ice is incompressible, and metal radiators are rigid. This stability can be exploited. For a car mechanic who investigates the breakdown, diagnosis is easy: if the night was cold, if there was water inside the radiator, and if the radiator is cracked, then that is the job done, epistemologically speaking. The mechanic may exploit a range of facts that were discovered in garages and laboratories elsewhere and that, it may safely be assumed, apply to this car here.

Label this case Stable Radiator. Now reimagine it to make it fragile. Suppose that, sometimes, water in radiators is mixed with antifreeze, cars are parked in high atmospheric pressure (so it is harder for water to freeze), or radiators are made from nonrigid material that can swell and bend. Any of these would break the connection between cold nights and cracked radiators. Suppose further that, in this new version of the case, it is hard to predict when these confounders occur and hard to confirm afterward that they have. Only extensive investigation can find out. Suppose that extensive investigation is also needed to check even whether the radiator has cracked.

Label this second version of the case Fragile Radiator. The connection between cold nights and cracked radiators, which previously was stable, now is fragile: it holds only unpredictably. This is true even though the underlying relations, such as that water expands when it freezes, remain as stable as ever. What matters is the relation *of interest*. For the mechanic, the relation of interest is that cold nights cause radiators to crack, and *this* relation is now fragile. It is not just lost in the noise of other factors; rather, the connection itself holds unpredictably.

Does fragility occur in real cases, not just in Fragile Radiator? Yes, all the time. These real cases include some of the most important challenges facing humanity: war, climate change, pandemics, economic inequality, elections, political violence, community and loneliness, ecological destruction, and many others. (The involvement of human agents in these cases is not what makes the difference, as we will see.) That covers the four horsemen of the apocalypse—war, pestilence, famine, and death—and plenty more besides.

---

[1] Hempel (1942).

And it is not just small details within these big questions; it is the big questions themselves. No single theory reliably explains all elections, all pandemics, all wars, or all aspects of inequality or climate change, nor can we discover without effort which theories explain individual cases. Fragility is everywhere.

Indeed, real cases are usually trickier than Fragile Radiator. The positive effect of a prospering economy on an incumbent's election chances, for example, is not stable. It can be disrupted by personal scandal, war, economic insecurity, immigration, voter prejudice against a candidate, pandemic, and no doubt by many other disruptors too: there is no complete list. Worse, the relations underlying these disruptions are themselves fragile. The impact of personal scandal, for example, varies idiosyncratically candidate by candidate and election by election. Unlike in Fragile Radiator, there is no neat taxonomy of disruptors and no known stable relations underpinning them. The scientist's task is daunting.

Fragility is not noise or messiness, which are properties of situations, not relations. Rather, fragility is a cause of difficulty in the face of noise or messiness: while stability guarantees us one reliable tool in the fog, fragility leaves us unsure of anything in advance of tentative local probing.

1.3 The impact of fragility

Fragility matters. It leaves science full of doubt and uncertainty and, with no theory to rely on, hostage to controversial local details. The public image and expectation of science—that it be decisive and impartial—are challenged.

To see why, and what is new about this, first consider an objection. Even in a stable world, relations are rarely universal, so in any given case we must identify which relations are operating. In Stable Radiator, for example, there is a menu of possible causes of car breakdowns. Yes, local investigation is needed to discover the actual cause, but no one ever denied that or denied that there is uncertainty beforehand. So, the objection concludes, why this big fuss suddenly about uncertainty and doubt and local detail? They were all in the picture already.

But this objection misses something important, and what it misses forms the heart of this book.

In one way, the difference between stability and fragility is indeed a matter only of degree. Relations sometimes hold and sometimes don't, and the only thing that varies is how predictably they do or don't. And yet, a switch from stability to fragility leads to a qualitative change methodologically. That is why it is important.

In Stable Radiator, the cold-radiator connection can be *predicted* easily: if there is water and cold, then the radiator will crack. But in Fragile Radiator, even with water and cold, it cannot be predicted easily, because sometimes there is antifreeze in the water, or the atmospheric pressure is high, or the radiator is flexible. Nor is that all. In Stable Radiator, the cold-radiator explanation can also be *confirmed* easily: just establish water and cold, and check that the radiator is indeed cracked. But to confirm the same explanation in Fragile Radiator is difficult: it is intricate business to establish that in this instance there was no antifreeze, the atmospheric pressure was normal, and the radiator was rigid. Lastly, in Stable Radiator, effective *interventions* are licensed easily: empty the radiator of water overnight and it will not crack, replace the cracked radiator and the car will work again, and so on. This effectiveness is why we go to the car mechanic in the first place. It follows from the relative ease of prediction and confirmation. But intervention is much less reliable in Fragile Radiator: further investigation is needed to know whether emptying the radiator of water overnight will solve the problem.

In Stable Radiator, cold nights crack radiators reliably. This is based on the stable facts that water does not compress, that water expands when it freezes, and that metal radiators are rigid. In turn, our belief in these facts' stability is based on the success of many previous interventions that rest on them and on coherence with wider theory. It is this that underwrites a car mechanic's interventions and explains those interventions' success. But in Fragile Radiator, cold nights crack radiators only unpredictably. The mechanic is forced each time, like Sherlock Holmes, to ferret out the details of the individual case, and the balance of effort shifts toward investigation of local nuance—so to speak, from the theorist to the case worker.

To get a sense of the consequences of this, consider elections again. Because no single theory holds reliably, usually there are many plausible explanations of an election result: the economy, a weak candidate, fear of immigration, and so on. Which explanation is right? To find out, we must delve deeper: is there supplementary evidence, such as interviews with voters or cross-analysis of vote shares in different districts, that supports, say, an economic explanation? This kind of case-specific inquiry is crucial. But it is sensitive to local detail, and so the explanations that it establishes do not extrapolate easily. Their warrant is *narrow scope*. Just because fear of immigration turns out to explain one election result, for example, that does not mean it also explains another.

Such explanations are often far removed from any wide-scope parent theory; instead, they are derived from some narrow-scope entity such as a bespoke causal model or a historian's single-case analysis. Even when explanations are derived from a parent theory, if relations are fragile, such explanations' warrant is typically narrow scope even if the parent theory is wide scope. When that is so, the truth of theories no longer matters. What matters instead is the truth only of the eventual narrow-scope explanations. Theory plays an instrumental role as an aid to developing these explanations. It might alert us, for example, to look for certain economic concerns of voters, such as job insecurity or wage stagnation in the middle class. But it is hard to know in advance the relevance of these concerns, and matters can be settled only by painstaking local investigation. A job for the case worker, not the theorist.

The more that relations are fragile, the more that science is not grand convergence on a few theories but instead is the worthy mastering of individual problems, one by one.

These methodological lessons carry bite. They tell against *laboratory experiments* whenever the laboratory is not itself the target context—and it almost never is. Laboratory sciences tend to investigate stable relations, and for good reason: a relation established in the laboratory may then be relied upon to apply outside the laboratory, too. In a stable world, laboratory methods are the route to success. But in a fragile world, they are not. For example, psychology experiments that investigate the Prisoner's Dilemma game are reliable guides to behavior only in laboratories, and even there not always. The methodological lessons also tell against *statistical analyses*—the mainstay of quantitative social science—whenever a statistical sample is not itself the target. For example, a statistical analysis of elections does not bear on an election here and now if the analysis' sample includes elections from too long ago, of too different a type, or from too different a country. Perhaps most significantly, the lessons tell against theory *developed in the abstract*. Theory developed like that, divorced from empirical refinement, plays its instrumental role badly: it is rarely an efficient aid for developing explanations. Yet theory is developed in this way often. The Prisoner's Dilemma game is an example, as we will see.

When relations are stable, these widespread methods are effective, even essential. That is why they are widespread. But when relations are fragile, these methods are a wrong turning.

Fragility has implications not just for science but also for my own discipline, philosophy of science. When I started, from a background in history and economics, I was caught off guard. The focus was almost always on famous theories from physics. Whether scientific theories are literally true, whether they are falsifiable, whether other sciences may be reduced to physics, Kuhnian revolutions, formal inductive logic, what laws of nature are—I loved all that, but it does not speak much to history or economics. Outside the guild, this reaction is common: is there a philosophy of science that speaks to *us*?

Philosophy of science has developed greatly in recent decades and is now a broad church. As will become evident, this book is indebted to excellent work on idealization, modeling, explanation, causal inference, and social organization. Yet even today, most examples in this work are examples of stability. And the same is true of a focus on mechanisms and even of work that encompasses "messy" aspects of science, such as uncertainty and stochastic systems, disunity, and scientific understanding. Comparatively little philosophy of science speaks to fragility.

1.4 One more heave?

But is this all too pessimistic? Rather than adapt to fragility, it might be objected, why not beat it—by finding new theories? We should persist, not just give up.

This optimistic rallying cry is inspired by the history of science. A great human achievement was the discovery in the seventeenth century of the differential equation that describes the motion of a pendulum. Some argue that the combination behind this discovery—calculus and experiment—was the intellectual engine behind the lift-off of modern science. Over subsequent centuries, this differential equation has reappeared whenever oscillations occur, whether it be "the worrisome movements of a footbridge, the bouncing of a car with mushy shock absorbers, the thumping of a washing machine with an unbalanced load, the fluttering of venetian blinds in a gentle breeze, the rumbling of the earth in the aftershock of an earthquake, [or] the sixty-cycle hum of fluorescent lights."[2] A single pattern of structural dependence, captured by mathematics, is stable across innumerable phenomena. It is a beautiful thing.

Similar examples abound. When relations are stable, traditional methods thrive: we can develop, test, and apply wide-scope theories, after the pattern of Stable Radiator. This strategy has reaped humanity a spectacular harvest. Why not stick with it? Who will discover, say, a differential equation that

---

[2] Strogatz (2020, 73).

accurately describes elections and fulfill Hume's ambition by becoming the Newton of the social sciences? That is the way to lasting progress, this objection concludes, not settling meekly for a fragile second best. We can get there. One more heave.

A bet on such deliverance is, in effect, a bet that the relevant stable relations exist. But there is no guarantee they do. Perhaps election results are indeed governed by as-yet-undiscovered stable relations—but perhaps they are not. What is the evidence? So far, not encouraging. It's not as if people haven't looked. Efforts to discover stable relations in the election case have persistently failed, and there is good reason to think they will continue to fail. To insist otherwise amounts to no more than an act of faith, unsupported by any trend in the science. Often, a relation is not easily rendered predictable just by learning more, and instead, we must take fragility as given. We will see this repeatedly throughout the book.

It is natural to seek whatever stability we can, but this can be a dangerous path. If relations are stable, we need methods appropriate to that; but if relations are fragile, we need methods appropriate to *that*. Just adopting stability-appropriate methods cannot magically create stability any more than any intervention on a symptom can change a cause. Bad science follows. Local investigations are neglected; ineffective theory is promoted. The aspect of this book that is critical is directed at those parts of science that make this mistake and at those parts of philosophy of science that defend it.

Might we veer away from stable relations too much, out of concern for fragility? Experience suggests the opposite. We will see again and again how the error is the other way around: stable relations are postulated or pursued when they should not be. Humans rarely need a second invitation to see patterns. Further, in those cases in which stability can be found despite initially unpromising circumstances, the best way to find it is via *fragility*-appropriate methods. The risk of missing out, ironically, is *increased* by one more heave.

There is a second version of one more heave. This time, the deus ex machina is digital: rather than by new theory, fragility will be overcome by algorithms and data—we will be able, it is hoped, to predict complex phenomena that currently we cannot, such as election results. But big data optimism should be tempered. For accurate algorithmic prediction requires stable correlations, and the evidence is that in many of the cases we care about—including those horsemen of the apocalypse—this stability is just not there, and nor will it likely be in the future. Again, to insist otherwise amounts to no more than an act of faith, and not a rational one.

1.5 Chapter summary of the book

1.5.1 Chapter 2: Definition of fragility

Roughly, a relation is fragile when its operation is unpredictable, as when, in Fragile Radiator, we don't know exactly when cold air will cause a cracked radiator. (The definition in the chapter is more precise.) Fragility is Janus-faced, because predictability is both objective and subjective, strongly influenced by external reality but also agent relative. Agent relativity is a feature, not a bug: optimal methodology *should* be relativized to investigator knowledge if our aim is to improve science as practiced by humans rather than by omniscient gods. Yet fragility is not arbitrary. Duly relativized, it is thereafter perfectly objective. And it is fragility's objective side that, in many cases, makes its methodological consequences inescapable.

Armed with a definition, we can organize the conceptual landscape. We see how fragility relates to philosophical theories of causation and explanatory power, how it relates to other concepts too, and why its implications are not hostage to various controversies in metaphysics.

1.5.2 Chapter 3: Going local: narrow-scope explanations

In Chapters 3 and 4, I turn to fragility's methodological implications. In a slogan, these implications are localist. I begin Chapter 3 with the first of several case studies in the book: the spontaneous truces that developed on the Western Front in World War One. These truces are explained only by local historical work, not by the general Prisoner's Dilemma game.

With stability, a theory established elsewhere may be assumed to hold here. For example, even if Coulomb's Law predicts badly because its effects are outweighed by other forces, still we believe that charged particles obey it. Why? Because Coulomb's Law has been vindicated by laboratory experiments, and we are confident that it holds stably. With fragility, though, vindication may no longer be imported from elsewhere. We always need fresh vindication here and now. That means empirical accuracy must be prioritized continuously.

With fragility, life is therefore more difficult: in noisy environments, empirical accuracy is difficult to achieve, which makes it hard to know which theories apply. For the most part, the only solution is supplementary, contextual investigations. These usually yield explanations whose warrant is narrow scope, that is, whose warrant covers only one or a few cases. Why? Because warrant requires empirical accuracy continuously, and noise means that to achieve empirical accuracy we must take account of local quirks—which implies narrow scope. Warrant for interventions and policy advice is usually narrow scope too, for the same reason.

1.5.3   Chapter 4: The core divide: Stability-Theorist versus Case-Worker

Chapter 4 is the central chapter of the book. In it, I set out the contrast between two methodological strategies, which I label Stability-Theorist and Case-Worker.

First, some preparatory groundwork. It is a truism that theories and background knowledge are essential even to local investigations. How may that be squared with localism? The answer is that the warrant for an explanation drawn from a wide-scope theory may be only narrow scope—if that wide-scope theory holds unreliably. Moreover, while sometimes explanations are derived directly from a parent theory, other times they emerge from a mélange of many ingredients, of which a parent theory is just one, which again generates explanations whose warrant is narrow scope. Generally, we draw from a large toolbox of theories, never knowing in advance which, if any, theories apply. The truth of a theory does not matter, only the truth of an eventual explanation. A theory's contribution to an explanation may be just heuristic. A case study of economic auctions illustrates this last possibility: no theory helped, except heuristically.

This leads to the dichotomy at the heart of fragility. Theory may be applied in two different ways. First, if relations are stable, a *Stability-Theorist* strategy is favored: apply a master theory, confident that it holds, and then fill in local details each time as required. The master theory is a reliable foundation. Second, by contrast, if relations are fragile, nothing may be assumed in advance. Now, it needs to be established each time which (if any) theory applies, and scientific effort must shift toward contextual investigations. I call this second strategy *Case-Worker*.

A paradigm of Stability-Theorist in action is the Newtonian two-body model of gravitational attraction. This model may be assumed to apply universally. All that is required each time is to fill in parameter values. But suppose instead that, unpredictably, gravitational attraction sometimes occurred and sometimes did not, or that, unpredictably, sometimes it followed an inverse-square law and sometimes an inverse-cube one. Now, each time, contextual investigation would be required to know what model applied. Perhaps, as with the economic auctions, each time a bespoke model would need to be created—a case for Case-Worker.

In noisy field environments, we frequently face a crucial methodological choice: should we insist on empirical accuracy, or should we let that slide, instead giving leeway to theories even when they predict badly? Many models in field sciences do predict badly—should we stick with them, nevertheless? The answer turns on fragility. If relations are stable, then Stability-Theorist works well, and sticking with a theory is justified, because we may be confident it is operating underneath the surface even when noise muddies the empirical water. This, advocates claim, is even *superior* to chasing empirical accuracy. Why? Because empirical accuracy in a noisy environment requires us to take account of every transient thing, but science should be in the business of abstracting away from the transient and tracking the seminal. But if relations are fragile, everything changes—now, unless we have empirical accuracy, we may not claim to have tracked anything. So, we can no longer afford to let empirical accuracy slide. Case-Worker is needed.

This core methodological dichotomy between Stability-Theorist and Case-Worker is perennial. It arises whenever it is hard to achieve generality and empirical accuracy at the same time—how, in such circumstances, should we theorize? What is new in the book is not the dichotomy itself but the response to it.

Much turns on this dichotomy. When relations are fragile, Case-Worker argues for the central importance of case studies as the best means to develop a theory and as the best means to establish indicators of when a theory applies. Case-Worker also endorses middle-range theories as the most likely to be useful.

This leads to the second appearance of the Stability-Theorist/Case-Worker dichotomy, now not with respect to theory application but with respect to theory *development*. If relations are fragile, theories should be developed according to Case-Worker. That means theory should not be developed in the abstract, perhaps mathematically, with wide-scope ambitions. A case study illustrates how theory should be developed instead: Donatella della Porta's work on political violence. Her theory is constantly refined by close empirical engagement. As well as establish explanations, it does what, in a fragile world, is a large part of a theory's job: it develops useful mechanisms and concepts—new tools for our toolbox.

Nonempirical theory development, abstracted from continuous empirical refinement, can work well when relations are stable. That is the

Stability-Theorist alternative. The problem is that many theories are developed in a Stability-Theorist way even though the domains they address are thick with fragility.

Next, I address two worries. First, how can we tell whether a relation is fragile? Answer: in practice, fragility can usually be discerned straightforwardly, so this is not a problem. Second, might prematurely adopting Case-Worker cause us to miss out on stability? But this worry, too, is misplaced. The far more common danger is adopting Stability-Theorist when Case-Worker is called for, not the other way around. And ironically, in any case, the best way to find pockets of stability in fields of noise is via Case-Worker, not Stability-Theorist.

The methodological ideas in Chapters 3 and 4 draw from and overlap with much work by others. In the penultimate section, I outline this debt explicitly. I relate the program of the book to recent philosophy of science, especially the work of Nancy Cartwright. There are many similarities. The main difference is that whereas Cartwright's system is built around an ontology of stable causal capacities, in this book, I explore the far-reaching consequences of, so to speak, capacities (and relations more broadly) instead being fragile.

I finish the chapter with a summary.

1.5.4 Chapter 5: Ubiquity of fragility

The preceding matters only if fragility is common. In Chapter 5, I show that it is—indeed, ubiquitous.

I begin with another case study—invasive species—to illustrate that fragility arises in natural science, too, not just in social science. Next, I survey a priori arguments offered by several authors for, in effect, why we should expect fragility to be widespread, especially in biological and social sciences. I then discuss two other general considerations. First, contrastive explanation: for any actual event, there are many contrasts relative to which the explanation of the actual event must appeal to a relation that is fragile. In this sense, fragility is ubiquitous. (There will also be many other contrasts that require appeal to a relation that is stable, so stability too is ubiquitous.) Second, external validity and extrapolation: concern about these directly implies fragility, concern about these is widespread; therefore, fragility is widespread too.

In addition to these general considerations, numerous examples show that whether a relation of interest holds often cannot be predicted easily. That is, fragility is widespread. I go through examples from environmental management, civil engineering, biology, mechanical engineering, medicine, cognitive science, criminology, political science, economics, and forecasting. Elsewhere in the book, I present several case studies. But it is also valuable to cover a greater number of examples more quickly, as in this section, to make it plausible that the longer case studies are not atypical outliers.

1.5.5 Chapter 6: Fragility and philosophy of science

In the remaining chapters, the book pivots from analyzing fragility to applying it. To begin, in Chapter 6: how does fragility bear on some debates in philosophy of science?

There are famous and effective solutions to the problem of noise: laboratory experiments, field trials, and many statistical methods. But these solutions all require stability. Without that, extrapolation of an experiment's results becomes dubious, and so does assuming a steady signal in the noise for statistical methods to detect. The value of laboratory experiments and many statistical methods is thus threatened by fragility, and it is no coincidence that laboratory sciences tend to deal with relations that are stable. Statistical methods, meanwhile, are most valuable in cases of stability plus noise.

Qualitative methods are highly useful and often essential. Case-Worker endorses them. This endorsement is now on the back of a methodology, that is, Case-Worker, that does not turn its back on third-person causal explanation or on prediction and that is not restricted to social science.

Fragility bears on scientific realism. In a fragile world, what matters is realism about narrow-scope explanations, not wide-scope theories. Further, instrumentalism about a wide-scope theory is viable only if that theory describes relations that are stable. Thus, the classic realist–instrumentalism debate applies only to half of science, as it were—the stable half. This bears, in turn, on the notion of scientific progress: when relations are fragile, progress is much more than better theories.

A fragility perspective bears on scientific explanation too. It tells against the unificationist view. But it endorses many aspects of a mechanistic view— although with a caveat, namely that we must avoid developing mechanisms in a mistakenly Stability-Theorist manner. How-possibly explanations, meanwhile, while sometimes useful, are also sometimes abused to excuse inappropriate Stability-Theorist work.

1.5.6 Chapter 7: Fragility and reflexivity

Reflexivity is, roughly, when to study or theorize about a target itself influences that target. Many take this phenomenon to be a significant methodological challenge and, further, a challenge unique to social science. Hence, social science must be fundamentally different from natural science, methodologically speaking. I disagree. I systematically go through cases that do and do not feature each of reflexivity and fragility, to show that fragility is what matters.

I then discuss how a focus on reflexivity can both help and hinder scientific progress. Finally, I return to the relation between natural and social science and how this looks from a fragility perspective.

1.5.7 Chapter 8: Fragility and economics

If theories are not to be judged by their truth, or by their closeness to the truth, then how should they be judged? The answer is by how well they play their toolbox role: do they help generate explanations that are true, and predictions and interventions that are accurate?

A toolbox role means that theories lose exclusivity. This raises an efficiency question: which theories, or other aspects of science, deserve the most resources? I examine critically the dominance (recently under pressure) of neoclassical orthodoxy in economics. I argue that fragility is the best tool for diagnosing economic theory's weaknesses and strengths.

I then extend the discussion to social science generally. I conclude that in social science, Newton-style grand theorizing is a mistake; it is a poster child for following Stability-Theorist inappropriately.

1.5.8 Chapter 9: Fragility and big data

Will big data make fragility yesterday's problem? Might number-crunching power sweep away our dependence on theory, fragile and stable alike, and usher in a new world: priority for prediction over explanation or causal understanding?

To reach this utopia, we must achieve accurate predictions. I examine four important phenomena: political elections, the weather, gross domestic product, and the results of interventions suggested by economic experiments. These examples suggest caution. Although big data methods are indeed very useful sometimes, in these examples, they improve predictions either limitedly or not at all, and their prospects of doing so in the future are limited too. I discuss what determines when big data methods succeed. A major constraint turns out to be fragility. It cannot be ignored or transcended, after all, because, roughly, it often precludes the stable correlations needed for big data methods to succeed (and even when it doesn't, it often precludes us from discovering them).

1.5.9   Chapter 10: Fragility and the COVID-19 pandemic

I examine in detail two examples of epidemiological modeling from the COVID-19 pandemic. The first is the model from Imperial College London that strongly influenced policy at a critical moment in March 2020, including the UK government's decision to impose a national lockdown. I conclude that this model fails: its policy recommendations carry no weight. The underlying reason is that its method is inappropriately Stability-Theorist. The second example is the estimate of the transmissibility of the Alpha variant made in December 2020, again in work from Imperial College London. This time, the work follows Case-Worker methods, and because of that it is successful.

1.5.10   Chapter 11: Conclusion: expertise in a fragile world

First, I summarize the book. Then, I briefly illustrate its themes with an example from actuarial science: two kinds of insurance underwriting instantiate the core divide between Case-Worker and Stability-Theorist, corresponding to whether the relations in actuarial models are fragile or stable.

In the rest of the chapter, I turn to what, in a fragile world, we should expect from an expert. Is there a crisis of expertise? If expertise is properly understood, no, there is not. Expertise has often been identified with knowledge of theory or methods. Case-Worker implies something different: knowledge of theory and methods, yes, but also substantive local knowledge, skill at identifying what knowledge is needed, skill at getting it, and then skill at using it. Such expertise is often only local. It implies a certain kind of practical proficiency, beyond that which can be written down or formalized easily in textbooks. It overlaps with views of expertise already developed in virtue epistemology, philosophy of science, and social epistemology.

Lastly, I ask briefly why Stability-Theorist is sometimes pursued erroneously and, thus, what lessons this book carries for different audiences.

1.6 What this book is and is not

This book is a work of general philosophy of science. I believe there are interesting and useful things to examine, at a level general enough to cut across many sciences, that philosophers are experts about and that many practitioners are not. Fragility is such a thing.

Like most works in general philosophy of science, this one is grounded in the sciences that its author knows best. For me, those sciences are economics and history. This background bears on the book's agenda. An immensely important issue in social science is the scope and use of empirical generalizations: methodological battles have been fought about this since the nineteenth century between, roughly, theorists and empiricists. But the issue extends well beyond social science. As we will see, it is salient in most sciences. General philosophy of science thereby benefits from a sensibility formed in philosophy of social science.

The book is aimed not just at philosophers of science but also at practitioners; at other philosophers; at other students of science; and at science funders, planners, and educators. Some technical machinery is required. But, especially in Chapters 7–11, which apply this machinery, the book is written to be widely accessible.

In some places, I go against some practitioners about methods in their own domains. But the book is not anti-science. I equally come out in favor of other, rival practitioners: often, there is methodological disagreement *within* science, and philosophers should not pretend they can avoid taking a stand. Many scientists do outstanding and difficult work in fragility cases. Philosophers should identify this work, figure out how it is done, and celebrate and promote it. The goal is to understand and enhance.

The two main theses in this book are:

1) When relations are fragile, there should be a shift from theorist to case worker.
2) Many relations of interest are indeed fragile.

I use the master idea of fragility to systematize a great many methodological reflections from many scholars and so to systematize a view of science. Fragility is a unifying thread for what otherwise might be articulated only in bits and pieces.

The achievements of stability science are awesome. Philosophy of science is admirably penetrating about them, and I teach that every year in my own classes. But much of science, I have come to realize, and much of the science that faces the public is dominated by fragility. We need a philosophy of science for that too.

2

Definition of fragility

2.1 Definition

In this chapter, we "eat our greens" by developing the definition of fragility used in the rest of the book. "Fragility," as I will define it, is a term of art. It sometimes departs from the word's everyday connotations and from previous uses of the word in philosophy.[1]

All relations hold in some circumstances but not in others. The following, for example, are true sometimes but not always: a cold night cracks a car radiator, a rise in national prosperity causes an incumbent to win an election, an open auction increases the price, a target being a small island causes a species invasion to succeed, and striking a match causes a flame. An exception is universal laws, which do hold always. But these are in short supply—to say the least—in the examples that we will be looking at. Another exception is idealized but false relations, which are widespread in science, and which, literally speaking, hold in no circumstances. With them, the salient point is that they are useful, perhaps approximately true, in some circumstances but not in others.

This fact—that relations hold sometimes but not always—raises a puzzle. Intuitively, fragile and stable relations differ objectively: stable ones hold more widely. But we cannot usefully measure "how many" circumstances different relations hold in, nor "how often" they hold, because the answer is the same for all relations: the number of circumstances they potentially hold in is indefinitely many, as is the number of circumstances in which they do not.

To make progress, we must go contextual. A parallel is with causes and background conditions, the difference between which, it is usually thought, cannot be defined purely metaphysically. Indeed, causal relations are the relations most discussed in this book, and it is well known that many properties of causal relations are sensitive to context.[2] It is hardly surprising that fragility is one of them. (Note: fragility is a property of relations generally, not just causal ones.)

---

[1] I have in mind especially the titles of Wilson (2004, 2005) and also fragility as a disposition of objects such as vases.

[2] Reiss (2013a, 2013b); Schaffer (2012); Northcott (2008a).

Science for a Fragile World. Robert Northcott, Oxford University Press. © Robert Northcott 2025.
DOI: 10.1093/9780191944352.003.0002

The distinction between fragility and stability is methodologically important, and we need to formulate it so that this importance is brought out. Here is a first, provisional definition: a stable relation is one that *holds across all salient circumstances*; a fragile relation is one that does not. But this is still not quite right. To see why not, tweak the Fragile Radiator example from Chapter 1: suppose in the new version, it is known that red cars—but not others—have rigid radiators, no antifreeze in the water, and are parked in normal atmospheric pressures. (Throughout this chapter, I use examples that are somewhat schematic, for clarity. There will be plenty of real-world examples in later chapters.) Cold nights therefore crack the radiators of red cars but not those of other cars. Label this version of the case Color Radiator. No need now to redirect scientific effort toward investigation of each individual case, as there was in Fragile Radiator. Rather, a mechanic can instantly tell just from the car's color whether the relation between cold nights and cracked radiators applies, even though the relation between cold nights and cracked radiators does not hold across all cases. Our provisional definition of fragility fails. The color clue makes all the difference.

Could we rescue the provisional definition by stipulating that there are now two different salient ranges of circumstances: red cars and other cars? Then, the cold night–cracked radiator relation would hold (or not) constantly within each of these subcategories, and the "holds across all salient circumstances" definition of stability would be satisfied after all. One obvious worry is ad hocness: what constrains us from defining "all salient circumstances" so as always to preserve constancy within any range of interest? Every relation could be classified as stable, and fragility would be defined out of existence. That is indeed a danger. But the real problem is something deeper: that the proposed rescue would move the definition away from what matters methodologically.

To see why, note that everything depends in our example on whether the mechanic knows the color code. If they do, then, as it were, stability methodology is appropriate; but if they do not, then it is not. What the agent knows is crucial, and it needs to inform a definition of fragility if we want that definition in turn to inform us methodologically. Agent relativity is required.

The key criterion is *predictability*. A relation may hold only intermittently, but if it can be predicted easily when it does and does not hold, then we are still in the methodological world of the original Stable Radiator, in which knowledge of wide-scope theories is king. That changes only if predicting when the relation holds becomes difficult. Only then, as in Fragile Radiator, are we forced to become like the fictional detective Sherlock Holmes, redirecting our efforts toward case-specific investigations.

One element from the provisional definition does survive, though: saliency. In any given case, the vital thing is not whether a relation holds predictably always but whether it holds predictably when it matters. Never mind whether, say, cold nights predictably crack radiators when antifreeze and flexible radiators have never been invented; what matters to a mechanic is whether they predictably crack the radiators of the cars the mechanic is treating here and now.

There is more to say. Predictability is a matter of degree: even if, in binary fashion, a relation either does or does not hold, whether it holds may still be predictable to a greater or lesser extent. True, predictability itself is sometimes interpreted informally as a binary, yes-no property. But taking predictability to be a matter of degree is supported by scientific practice, as in fields such as weather and political forecasting—even the best predictions in science are never completely certain.[3]

Some notes on terminology: first, predictability here may be understood retrospectively as well as prospectively. For explanatory purposes, it matters whether we can be sure a relation *did* hold, just as for predictive or planning purposes it matters whether it *will* hold. Second, I will use "reliable" as a synonym for "predictable." And third, I will use a relation "holding" as a synonym for a relation "operating."

Turn now to fragility itself, as opposed to predictability. In colloquial usage, fragility too is something that can be understood either as a binary or as a matter of degree. But for our purposes, stipulate that fragility is a binary, yes-no property: a relation either is or is not fragile. Why? Because methodologically, there is a dichotomy between two strategies, roughly, between that of the theorist and that of the case worker. I return to this dichotomy in detail in Chapter 4, where I label the two strategies *Stability-Theorist* and *Case-Worker*. For now, briefly: with Stability-Theorist, a theory is developed and applied in full confidence that we know when it will and will not operate. A paradigm of Stability-Theorist is the development and use of Newtonian theory, whose basic elements can be assumed to apply universally. With Case-Worker, by contrast, things are less reliable. Whether a theory applies cannot be safely assumed in advance, so which theories do and do not apply must instead be established afresh with each application. This deeply impacts, too, how theory is best developed, as we will see in Chapter 4.

For defining fragility, what matters is whether a relation holds predictably enough to favor the Stability-Theorist strategy or *un*predictably enough to

---

[3] Ghomi (2022).

favor the Case-Worker strategy instead. If the former, the relation counts as stable; if the latter, then fragile. *How* unpredictably must a relation hold to recommend the Case-Worker strategy? That depends on the details of the case: on the circumstances in which the relation is operating and on the purpose of an investigation. This is another way in which fragility is a contextual property.

The definition of fragility is therefore intricate. Fragility is defined in terms of predictability, yet while predictability is a matter of degree, fragility is dichotomous. Whether the unpredictability is enough to deem a relation fragile depends on whether it is enough to favor the Case-Worker rather than Stability-Theorist strategy, but the exact tipping point is contextual.

Putting all this together, define fragility as follows:

Definition
*A relation is **fragile** if and only if, in the salient range of circumstances, it does not hold predictably enough to favor the Stability-Theorist rather than Case-Worker strategy.*

Notes:

1. Stability is defined reciprocally: a relation is *stable* if and only if, in the salient range of circumstances, it *does* hold predictably enough.
2. A relation can be fragile at either the type or token levels.
3. "Salient circumstances" may include counterfactual ones.

Strictly speaking, fragility, so defined, is redundant. There are two axes of variation, predictability (continuous) and methodological strategy (dichotomous), covered by three label pairs: unpredictable versus predictable, fragile versus stable, and Case-Worker versus Stability-Theorist. Hence, the redundancy. Without loss of meaning, this book could be retitled "Science for an unpredictable-enough-that-Case-Worker-is-favored world." But it proves very helpful to have a label specifically for when unpredictability *is* enough to favor the Case-Worker strategy, because this connects a relation's unpredictability to the core methodological issue. Hence, fragility. It is a marker of whether Case-Worker or Stability-Theorist is favored, and this makes it a key guide to much else.

Predictability can often be improved by contextual investigation; indeed, for many purposes, it must be. In this sense, the operation even of fragile relations can be made predictable. But the point is the need for this extra effort: even after a detailed investigation in one context, a fragile relation's operation is still unpredictable as soon as we move to a new one. The contrast is between stable relations, whose operation can be relied upon across many contexts without the need for extra investigatory effort each time, and fragile relations, whose operation can*not* be so relied upon. This contrast is the one that matters methodologically.

Predictions in science are usually predictions of outcomes, not of whether a relation holds. True enough. But predicting an outcome requires the prior predictability of the relations behind that outcome—predicting that an object will fall depends on knowing that gravity will be working. So, fragility of relations is key.

We may represent the definition in a 2×2 table (Table 2.1):

Table 2.1 Definition of fragility

	Relation holds unpredictably	Relation holds predictably
Outcome is unpredictable	Fragile Case-Worker strategy	Stable Stability-Theorist strategy
Outcome is predictable	(Empty)	Stable Stability-Theorist strategy

In the two slots in the right-hand column, the relation holds predictably, Stability-Theorist is favored, and so the relation is stable. In the top-left slot, by contrast, the relation now holds unpredictably, Case-Worker is favored, and so the relation is fragile.

In the bottom-left slot, the outcome is predictable (by us). An outcome that is predictable by us would *be* predictable by us only because we had a means of predicting it, that is, we had a relation that holds reliably that we could use. But a relation that holds reliably would enable us to move to the bottom-right slot, which presumably is always preferable. Therefore, the bottom-left slot is always empty—or at least, I know of no cases in which it is not. Although it is hard to see the epistemic motivation for ever doing so, logically speaking, one could apply a relation that does not hold reliably even to an outcome that is predictable. In that case, one should follow Case-Worker, and so the relation would be fragile.

The Case-Worker and Stability-Theorist strategies are relevant only to relations in a theory or model—the relations that we use to do things like explain,

predict, intervene, and understand. It is these relations that are methodologically important. Do *they* hold predictably?

In the case of explanation, the relation that matters is the one we use to
do the explaining; that is, in the standard terminology, it is the *explanans*
relation. The outcome, meanwhile, is that which is predicted or explained
by the explanans relation; that is, in the standard terminology, it is the
explanandum.

As I will discuss in later chapters, a popular response to an unpredictable
outcome is to pursue Stability-Theorist and develop theory accordingly. This
makes perfect sense if we are in the top-right slot, where a reliable theory is
a welcome bulwark in the fog of noise. But if our theory itself holds only unpredictably, that makes all the difference; now, we are in the top-left slot, and
we must switch to Case-Worker. The heart of this book is to draw attention
to that.

The table also helps us understand the "one more heave" response to
fragility—the view that, when a relation does not hold predictably, rather than
settle for Case-Worker we should instead persevere with Stability-Theorist,
in the hope we eventually discover a relation that *does* hold predictably. "One
more heave" is a gamble—that in the table we will be able to move from the
left-hand column to the right-hand one. In the bottom row, when an outcome
is predictable, this gamble is justified. Matters are less certain in the top row,
though: starting from the top-left slot, will we be able to move to the top-right
one? True, there is no a priori reason we will not find a relation that holds
predictably—but equally, there is no a priori reason we *will*. We must judge the
prospects case by case. As we will see, often no stable relation is in prospect,
and then "one more heave" becomes damaging, as it foregoes Case-Worker for
nothing. This is the risk of always *assuming* that there are stable underlying
capacities or mechanisms waiting to be found: it pre-commits us to Stability-
Theorist, come what may.

Things would be simpler if all fragile relations were "one and done,"
unique to a time and place and never recurring. But reality is not like that.
Commonly, fragile relations are "in and out," holding multiple times but
not always, like the causal relation in Fragile Radiator between cold nights
and cracked radiators. Because stable and fragile relations may each hold
in multiple circumstances, so, as we saw, the distinction between them
cannot be how *widely* they hold. Rather, it is how *predictably* they hold.
But predictability is agent relative: knowledgeable agents can predict
better than ignorant ones. So, an element of subjectivity enters. Turn to
that next.

2.2 The two faces of fragility: objective and subjective

Predictability has two faces, thus, so too does fragility. Appreciating this is crucial for what is to come.

The first face is agent independent: some relations are easier to predict than others. In Fragile Radiator, some cars have antifreeze, flexible radiators, and diverse parking histories, while others do not. This extra variation is absent from Stable Radiator, which makes the cold night–radiator relation easier to predict—objectively. Similarly, whether a hurricane occurs can turn on the flap of a butterfly's wing halfway across the world, but whether a bridge remains standing is much less volatile, requiring only that there be no asteroid strikes, bombs, or other such extreme events. This contrast between hurricanes and bridges is objective. Indeed, it turns on the Earth's weather system being chaotic, and chaos can be defined objectively. As a rule of thumb, in complex, changing circumstances, predicting whether a relation holds is difficult more often and so fragility is more likely, whereas in simpler, less changeable circumstances, the opposite is true.

But predictability has a second face, too. In Color Radiator, whether a cold night will cause a cracked radiator is predictable if we know the color code but unpredictable if we do not. It is agent relative. Even the extreme butterfly-hurricane case presumably is predictable to an agent that, like Laplace's Demon, has superhuman knowledge of the atmosphere's every molecule.

Predictability is also agent relative in a further way. Suppose that expensive cars' colors reveal their radiator types, antifreeze, and parking history, but that cheaper cars' colors do not, so the situation is a mixture of Color Radiator (for expensive cars) and Fragile Radiator (for cheap cars). Next, suppose that a mechanic handles only expensive cars. In these circumstances, whether a cold night will cause a cracked radiator is predictable. But suppose that in different circumstances, the mechanic now handles only cheap cars, so matters are again unpredictable. Same underlying situation, same investigator knowledge, different circumstances salient—in one case, the mechanic handles only expensive cars; in the other, only cheap ones.

In sum, fragility is relative to the world (does the world make this relation easy to predict?), to agent knowledge (is this agent an expert?), and to agent interest (which circumstances are salient?). We might distinguish the latter two relativizations as being respectively epistemic and pragmatic, thus making fragility "three-faced." But I will subsume them under the single label "subjective."

Fragility is not arbitrary in any objectionable sense. Once relativized to investigator knowledge and interest, it is thereafter perfectly objective. And there are strong uniformities across all scientists and across all informed human agents generally, so that in practice there is rarely much uncertainty about whether a relation is fragile. For almost all relativizations ever likely to be relevant, for example, whether a butterfly's wing flaps cause a distant hurricane is fragile, whereas whether bridges stay standing is not.

The subjective face of fragility, meanwhile, is a feature not a bug. Indeed, its absence is a major gap in previous work. Optimal methodology *should* be relativized to an investigator's interests and to the knowledge that they could reasonably be expected to have: the norm is that appropriate for appraisal.[4]

Things can change with technology. In Fragile Radiator, for example, the Case-Worker strategy is favored, but suppose that a technological advance enables us to move to Color Radiator: a way is found to color-code each car's radiator, antifreeze, and parking history. Now, the Stability-Theorist strategy is favored. Car mechanics face the same task as before—to diagnose and repair radiator breakdowns—but the best way to perform that task is changed by the technological advance. Fragility's relativization to investigator knowledge enables us to capture this. Equally, things can change the other way. Technological regression or other loss of knowledge can change a relation from stable to fragile, as would happen if the color code were discontinued in Color Radiator or if a new mechanic did not know the code.

If fragility is relativized to investigator knowledge and level of technology, does this remotivate "one more heave": can fragility be overcome by new knowledge or technological advance? Sometimes it can, no doubt. But one more heave is still usually a mistake. The reason is fragility's objective face: some domains and focuses of interest are likely to remain thick with fragile relations indefinitely. For any plausible level of technology, for example, we will not be able to predict whether a butterfly's wing flap causes a distant hurricane. Nature is not always kind, and often fragility is here to stay.

Fragility's various relativizations enable us to accommodate two important features at once: whether a relation is fragile can change, and yet for practical purposes many relations are likely to remain fragile indefinitely, no matter what we do.

How does one *find out* if a relation is fragile? In reply: once one tries to use a relation, usually its fragility or stability becomes obvious soon enough, so in practice the issue is not pressing, as case studies show clearly. But I discuss that

---

[4] Steinberger (2019).

more later in the book (see especially Section 4.9). In the rest of this chapter, I focus on fragility's conceptual aspects.

2.3 Fragility versus mere contextual sensitivity

A fragile relation is usually sensitive to context, but not all contextually sensitive relations are fragile: sometimes a relation is sensitive to context, yet predictably so.

Consider the "law of demand," that is, that an increase (decrease) in economic demand for a good will increase (decrease) that good's price, and vice versa. This is sensitive to many background conditions.[5] There must exist a functioning market, that is, regular exchange of the relevant good; these exchanges must be regulated by appropriate social and legal norms; people must trust each other enough to overcome fear of fraud and informational asymmetries; people must have certain tastes; the existence of money is required—which in turn requires many supporting institutions; the supply of the good must remain stable enough; and there must be no price controls. Without any one of these, the law of demand breaks down.

The key point: these necessary background conditions are well known, and whether they hold can be easily observed, so whether the law of demand will operate can be well predicted without the need each time for contextual investigation. The law of demand therefore counts as stable. Of course, because the law of demand does not operate universally, sometimes we will need an alternative theory or model, but that does not vitiate its usefulness whenever the background conditions do hold. What matters is *not* whether a relation is "stable" in the sense of not varying much with background conditions, but instead whether it is stable in this book's sense of holding predictably.

Such "stable-contextual" relations, as we might call them, are widespread. First, there is a sense in which any relation is *potentially* stable-contextual—if only we knew their triggering conditions and when they held. So, no fragile relations after all? But often, we do not know the triggering conditions or do not know when they hold, and there is no prospect of doing so, so dreams of one more heave are hopeless. Second, all stable relations hold only sometimes, so all are, strictly speaking, stable-contextual. The only exceptions are relations that hold universally.

---

[5] Reiss (2017).

2.4 Fragility and causation

In many of the cases we will look at, the models and explanations of interest are causal, and accordingly the main relations of interest are causal too. How does fragility relate to causation?

Following the science, I sometimes discuss singular and local causes and sometimes also discuss macro-level causes. Beyond the weak constraint of accommodating such discussion, in this book, I am not committed to any specific theory of causation. Here, I will use James Woodward's influential manipulationist account for illustration.[6]

According to Woodward, causation consists in a dependency relation between cause-and-effect variables that is invariant over some range of interventions on the cause. This dependency relation holds across some range of background conditions. The wider this range of background conditions, and the wider the range of interventions for which the relation is invariant, the more that a causal relation is, in Woodward's terminology, "insensitive" rather than "sensitive." Insensitivity here, like stability, is associated with a greater ability to control. This is because Woodward's notion of sensitivity is related to the objective face of fragility: if a causal relation holds over only a small range of interventions or background conditions, that is a source of fragility rooted in objective circumstances. Woodward also acknowledges one of the subjective aspects of fragility, namely the interest relativity of which circumstances are salient. Reflecting that his concerns are not the methodological ones of this book, though, Woodward rejects the other subjective aspect of fragility: causal sensitivity, on his account, is *not* relative to investigator knowledge.

What matters methodologically, according to me, is whether (in Woodward's terminology) we can predict where and when an invariant dependency relation holds—or, alternatively put, whether in any given case we can predict which such relation holds. In Stable Radiator, it is easy to predict when the invariance relation between cold nights and cracked radiators holds, and so that relation is stable. But in Fragile Radiator, it is not easy to predict, and so the relation is fragile.

Not only a causal relation can be fragile; so too, can its strength be. This is obscured in Stable Radiator, because the cold night simply does or does not crack the radiator. But in many cases, we are interested in matters of degree: *how many* extra votes for an incumbent does a booming economy cause, and *how much* does it increase the incumbent's probability of winning?

---

[6] Woodward (2003, 2006). My own preference is for a simpler counterfactual theory (Northcott 2021).

Defining degree, or strength, of causation is an intricate matter.[7] But on any plausible account, degree of causation is highly sensitive to changes in background conditions, which makes it a hard thing to predict—often far harder than the mere fact of causal connection itself.

Not only simple causal relations may be fragile. So too may be, derivatively, relations associated with more elaborate causal structures, such as mechanisms, pathways, and cascades.

Many noncausal relations are important to science, of course, such as nomological, structural, functional, compositional, and mathematical ones. All of these relations can be fragile or stable, just as causal ones can: fragility is a property of relations generally. But a danger lurks. Laws connote stability: they are usually cited just when some relation persists reliably across cases. Structural and functional dependencies also usually persist reliably across cases. And this can mislead. For if your only goal is to discover laws, then the Stability-Theorist strategy associated with stability might seem appropriate always—even when it is not. In Fragile Radiator, for example, although there are laws that relate cold nights to cracked radiators, the relation *that we are interested in*—between cold nights and cracked radiators—is fragile, and methodology must adapt accordingly.

2.5 Scope

Fragility is a property of relations *in* the world, modulo its relativizations. *Scope*, by contrast, is a property of descriptions *of* the world, such as theories, models, and explanations. These latter entities can vary on a spectrum from *narrow scope* to *wide scope*. By narrow scope, I mean something that applies only to a few cases, and in the limit only to one case, while by wide scope, I mean something that applies to many cases, and in the limit universally. Scope is similar to what Petri Ylikoski and Jaakko Kuorikoski call "nonsensitivity" and what Christopher Hitchcock and James Woodward call "explanatory depth."[8] In all of these, the core concern is the same: how widely does an explanation apply?

Scope is an important notion for this book. If a causal relation is stable, for example, then it holds reliably across many cases, and we can be confident that the associated causal explanation is wide scope. But if a causal relation is

---

[7] Kaiserman (2018); Northcott (2013b).
[8] Ylikoski and Kuorikoski (2010); Hitchcock and Woodward (2003).

fragile, then, at least in simple instances, it holds across only a few cases, and the associated causal explanation is narrow scope.

Matters become more complicated if a fragile relation holds sometimes here sometimes there, that is, widely but intermittently. The associated explanation is then itself true widely but intermittently, and so is "wide scope" in this inter-mittent sense. But because the underlying relation is fragile, we cannot assume that an explanation that is successful in one instance will remain successful in a new instance, and so its *warrant* is narrow scope and needs to be re-established each time. I will return to this crucial point more than once. Even when, with a hypothetical God's-eye view, we would have known where and when some model explains, if the relations in that model are fragile, then its successes cannot be known in advance by *us*.

To illustrate: suppose that, in Fragile Radiator, a mechanic establishes in one case that a cold night indeed caused a car radiator to crack. In other words, the mechanic established that there was no antifreeze in the water, the radiator was rigid, and so on. Next, a second broken-down car ar-rives. Is the same explanation true again? We need a new investigation to find out, because the same explanation cannot be *assumed* to apply to the second car as well, *even if* eventually, after further investigation, it turns out that it did.

When defining a model or explanation's scope, how do we individuate cases? By number of *actual* instances? But then unsuccessful explanations would have zero scope. By number of *possible* instances? But then all explanations, except impossible ones, would have indefinitely wide scope. The lesson: scope must be defined relative to some salient range and individuation of cases; in other words, defined relative to investigator interest. In practice, though, inter-subjective agreement is nearly always more than enough to anchor the main methodological lessons.

A model or explanation is often of wider scope when its language is more abstract.[9] Its usefulness in particular cases then turns on the usefulness of par-ticular realizations of its abstract terms.

Ceteris paribus laws are wide scope, as they are presumed to apply to many cases. So too are stable causal tendencies or capacities.

---

[9] Cartwright and Hardie (2012).

2.6 Relation to other concepts

Fragility is defined in terms of unpredictability, so it might seem that it can be subsumed into other concepts (as they are often used). But it cannot.

Fragility is not the same as *complexity*, although complexity can lead to it. Fragility is a property of relations, while complexity is a property of systems. That said, complexity has several different definitions, and if it is defined in terms of unpredictability, then any fragile relation will inevitably have arisen from some complex system or other. But on most definitions of complexity, fragile relations can arise in noncomplex systems too, as in the World War One truces example in Chapter 3. Conversely, on any of the definitions of complexity, some relations even in complex systems are not fragile. Summers are predictably warmer than winters, for example—that is, seasonality stably causes this variation in temperature, even though the weather system is a paradigm of complexity.

Might a fragile relation just be one that is *indeterministic*? True, indeterministic relations are unpredictable or at least predictable only probabilistically. But indeterminism is not necessary for fragility: the relation between cold nights and car radiators in Fragile Radiator is deterministic, yet fragile. The problem is agents' ignorance of whether the relation holds. And indeterminism is not sufficient for fragility, either: some indeterministic processes are predictable enough for many purposes, such as that radioactive decay poisons, that a casino profits from games of chance, or that a vaccine makes serious illness less likely.

At the heart of fragility is *ignorance* and *uncertainty* about when a relation operates. But to use these terms in place of fragility would invite confusion, as one can be ignorant of, or uncertain about, many things—even in stability cases, there may be much ignorance and uncertainty. For our purposes, what matters is only a particular kind of ignorance and uncertainty, namely the kind tracked by fragility.

If we do not know whether a relation applies, then we have *underdetermination* of theory (i.e., of which relation applies) by evidence. Is fragility nothing more than that? No. While underdetermination might be necessary for fragility, it is not sufficient. Underdetermination applies widely—indeed universally, according to Quine and others—and so it applies to cases of both stability and fragility alike. Only a subset of underdetermination cases features relations that are fragile.

Is fragility just high sensitivity to environment, that is, just strong *interactive effects*? Such interactivity does indeed capture the objective face of fragility. But it misses fragility's subjective side. If interactive effects are known, or if they are irrelevant to the purpose at hand, then it will be predictable whether, in the salient circumstances, a relation holds, which means no fragility. Interactive effects must themselves be both relevant and unpredictable to have the consequences for methodology that fragility tracks.

Is (lack of) fragility the same as *projectability*? A predicate or kind is projectable when past instances can be taken as guides to future ones, which implies a certain predictability in just our sense. Projectability can diverge from fragility, though: a kind may be projectable, but if we do not know that, then we would reasonably follow Case-Worker. Again, only the objective face of fragility is captured. Further, fragility is a property of a relation, not of a predicate or kind.

Is fragility the same as *contingency*? No. Exactly how best to understand contingency (in the context of special sciences rather than logic) is a contested issue. I will take contingency here to mean something like causal sensitivity: an outcome is contingent if it could easily have been different, and it is necessary if it could not.[10] Contingency, so understood, is a property of an outcome or a system, whereas fragility is a property of an individual relation. Contingency often *leads* to fragility, by increasing the epistemic requirements for predictability and thus for stability, but it does not entail it. The output of some intricate Newtonian system, for example, might be highly contingent on that system's details but nevertheless be best investigated by Stability-Theorist methods, because the system's component parts operate predictably. The reverse holds too. One can have fragility without contingency, as when an outcome is stably determined but we human investigators are unaware of the full causal story, so the outcome is still unpredictable to us, and Case-Worker methods are best. This is the subjective side of fragility.

Numerous other familiar dichotomies fail to track the fragility–stability divide satisfactorily, as we will see: field versus laboratory sciences, social versus natural sciences, interpretivist versus naturalist views of social science, holism versus individualism, mechanistic versus statistical causal inference, macro versus micro, token versus type, technology versus science, and applied research versus pure research. Fragility versus stability cross-cuts all of these.

At root, what matters methodologically is: does a relation operate reliably enough for us to follow the Stability-Theorist rather than Case-Worker strategy? Fragility is defined to latch precisely onto this. No other concept is.

---

[10] Ben-Menahem (1997, 2009).

2.7 Metaphysical foundations

Is fragility merely superficial? Although in Fragile Radiator it is fragile that cold nights cause cracked radiators, underneath that fragility there are perfectly stable physical laws that determine when radiators crack, how antifreeze works, and so on. If we worked out how these stable laws interact, perhaps the surface fragility could be overcome? This line of thought leads to another "one more heave" proposal, now with a reductionist flavor: sidestep fragility by focusing on stable relations underneath the surface froth. Precisely because of their stability, this proposal suggests, these underlying relations are the only genuine foundation for scientific progress.

Much science is motivated by exactly this hope. The impact of the economy on election results, for example, might be muddied by other factors, but maybe stable relations exist that underlie the surface electoral phenomena, just as physical laws underlie the surface phenomena in Fragile Radiator—perhaps stable relations that link prosperity and voter psychology? The hope is that fragility can be dissolved completely. If we identify a core of fundamental or natural relations that are stable,[11] and if fragile relations can always ultimately be reduced to combinations of these stable ones, then science will be liberated.

But this hope is illusory, as much philosophical work has established. For how would the reductionist dissolution work? First, we must discover the relevant underlying relations—in many cases, as we will see, this is not easy. Second, these underlying relations must be stable—but there is no guarantee they will be. Third, we must discover the reduction—how do the postulated stable underlying relations combine to generate the fragile surface one? And fourth, the underlying relations, even if discovered, must also combine to produce the higher-level relations in a stable way. Rarely will all four of these conditions be satisfied.

We can now see why fragility's methodological consequences are independent of a debate in metaphysics. That debate is between two positions:

(1) There are stable fundamental properties and relations. Fragility is only ever the result of interactions between those fundamentals, even if, in practice, these interactions can be too complex for us to discover.

(2) At least some fragile relations are irreducible to stable ones.

---

[11] For example, those relations that stem from "perfectly natural" properties in the sense of Lewis (1983).

The central claims of this book, recall, are that fragile relations are wide-spread and that a distinctive methodology is required to study them. The most popular versions of (1) are compatible with both of these claims, because these versions do not imply methodological or explanatory reductionism, only onto-logical reductionism. For these versions of (1), therefore, I am neutral between (1) and (2).

The take-home message: *no a priori escape route*. Fragility's methodological consequences cannot be sidestepped purely with metaphysics.

The methodological claims in this book are independent of another meta-physical debate too, namely that surrounding Humeanism. In our terms, the Humeanism debate amounts to, roughly, whether "natural," "fundamental," and the like are just labels we put on kinds or properties that form significant stable relations (Humean view), or whether these natural kinds or properties also *explain* that stability (non-Humean view). But fragility's methodological consequences are the same either way.

Finally, two further metaphysical questions. First, *why* are some relations fragile and some stable? Often, we have proximate answers: in Fragile Radiator, for example, the relation between cold nights and cracked radiators is fragile because antifreeze is present unpredictably, radiators are flexible unpredict-ably, and so on. But if we push further for ultimate answers, I have nothing to add. The reductionists, and the critics and defenders of Humeanism, do have something to add.

Second, does fragility bear on natural kinds? Typically, I argue in Chapter 5, the same kind partakes in some relations that are stable and some that are fra-gile, and it is a contingent matter, logically speaking, which we happen to focus on more. Perhaps natural kinds can be seen as those that, in salient circum-stances, tend to partake in relations that are stable. This would be in sympathy with pragmatist views of natural kinds.[12]

The message of this chapter overall: fragility as I define it plays a precise philosophical role. It zeroes in on a key methodological fault line, as a marker of when a relation's unpredictability dictates the Case-Worker rather than Stability-Theorist strategy. Doing so makes clear that fragility must be relativ-ized to agent knowledge and interests. It also makes clear how fragility relates to, yet is distinct from, various other concepts in philosophy of science.

---

[12] Dupré (1993); Magnus (2012); Khalidi (2013).

3

Going local: narrow-scope explanations

3.1 Example of going local: World War One truces

What follows methodologically from fragility? In this chapter, I begin to answer that. I start with a case study, the first of several case studies in the book. Why include them? While general analysis implicitly promises to shed light on individual cases, insight flows in the opposite direction too: individual cases illuminate general analysis.

In World War One, at many places on the Western front, there developed a live-and-let-live system of informal truces.[1] These truces could consist of complete nonaggression, temporary periods of nonaggression such as at mealtimes, certain areas of nonaggression such as mutually recognized safe areas, or other arrangements such as intricate local norms about what actions and responses were or were not acceptable. The truces arose spontaneously, despite constant pressure against them from senior commanders. What explains this remarkable and moving phenomenon?

To answer that, the political scientist Robert Axelrod applied a wide-scope model: the Prisoner's Dilemma game.[2] His analysis is widely lauded. But upon closer inspection, it reveals the weakness of the Stability-Theorist strategy (to use the term I develop in Chapter 4) when relations are fragile. For the relations described by the Prisoner's Dilemma do not hold predictably.

In the Prisoner's Dilemma game, two players each face a choice between "cooperating" and "defecting." How well they do depends both on their action and on their opponent's action. In the version of the game that Axelrod used, if both players cooperate, they each get a payoff of 3, but if both defect, they each get a payoff of 1 (see Table 3.1). So, better off cooperating? Not so quick. The twist is that if one player cooperates and the other defects, then the cooperator suffers a payoff of 0 while the defector enjoys a bonanza payoff of 5. Therefore, as Table 3.1 shows, each player is always better off defecting, regardless of what

---

[1] This section is adapted with permission from Northcott and Alexandrova (2015) and Northcott (2018). See those papers for further details and discussion.
[2] Axelrod (1984).

Science for a Fragile World. Robert Northcott, Oxford University Press. © Robert Northcott 2025.
DOI: 10.1093/9780191944352.003.0003

Table 3.1 Prisoner's Dilemma payoffs

	Cooperate	Defect
Cooperate	3,3	0,5
Defect	5,0	1,1

their opponent does. This is the "dilemma." If only players could agree to co-operate, they would each get a payoff of 3, but the strategic situation incentivizes each of them instead to defect, which traps them in the defect–defect outcome with a payoff of only 1.

Things change, though, if the game is *repeated* many times. Now (glossing over details), the promise of sustained gain in the future from mutual co-operation can outweigh the temptation of an immediate, one-off gain from defecting. Two players might build up trust, thereby enjoying time after time a payoff of 3 from mutual cooperation and avoiding the trap of mutual defection and a payoff of 1. The shadow of the future makes the difference.

Axelrod argues that the implicit payoffs facing the soldiers in the trenches formed just such a repeated Prisoner's Dilemma. To defect corresponds to shooting the enemy when you have the chance, while to cooperate corresponds to *not* shooting. The game is *repeated* because the situation across the trenches was stable over relatively long periods. The payoffs are from the soldiers' point of view, and it is assumed the soldiers preferred the safety of mutual no shooting, even if that harmed the overall war effort. (The soldiers' commanders, of course, might have had very different preferences.)

If the situation in the trenches was a repeated Prisoner's Dilemma, what behavior should we expect? According to Axelrod, the answer is the "Tit-For-Tat" strategy with initial cooperation: players initially play cooperate and thereafter repeat whatever the other player played in the previous period, whether that be cooperate or defect. If both sides play this strategy, the result is sustained mutual cooperation. This, Axelrod thinks, explains the truces.

The repeated Prisoner's Dilemma allows for many possible outcomes besides indefinite mutual cooperation. Why think players will follow Tit-for-Tat? Axelrod appeals to the results of his well-known computer tournaments, in which many game theorists submitted strategies, competing against each other to see who could achieve the highest payoffs in repeated Prisoner's Dilemma games. The winning strategy in these tournaments was Tit-for-Tat with initial cooperation. Therefore, Axelrod concludes, Tit-for-Tat is the strategy we should expect the soldiers to follow. I will use "Prisoner's Dilemma" as shorthand for this richer analysis of Axelrod's. The main lesson of this section, namely that Prisoner's Dilemma fails to explain, would apply still more strongly to the Prisoner's Dilemma game alone because then we would face the additional problem of justifying our selection of the mutual cooperation outcome.

Axelrod's explicit goal is explanation (italics added):

The main goal [of the case study] is to use the theory *to explain*:

1) How could the live-and-let-live system have gotten started?
2) How was it sustained?
3) Why did it break down toward the end of the war?
4) Why was it characteristic of trench warfare in World War I, but of few other wars?[3]

He draws on the fascinating and detailed account of World War One trench warfare by the historian Tony Ashworth.[4] Initially, historical details might seem to support Axelrod's case. Both sides demonstrated their force capability—but in harmless ways, such as by expertly shooting up a barn. Tit-for-Tat implies, according to Axelrod, that to demonstrate a credible threat is optimal but to actually defect is not optimal, so establishing credibility in a nonharmful manner should be expected. It also predicts that repetition is crucial: cooperation requires the Prisoner's Dilemma game to be played without a known endpoint. This too was borne out in the trenches, as old hands typically instructed newcomers carefully in the details of the local truce's norms, so those norms greatly outlasted the time any individual soldier spent on that part of the front.

Perhaps Axelrod's most striking evidence is how the live-and-let-live system eventually broke down. The unwitting cause of this, he argues, was a policy, dictated by senior command, of frequent raids, that is, of carefully prepared attacks on enemy trenches. If successful, prisoners would be taken; if not, casualties would be proof of the attempt. Axelrod comments:

There was no effective way to pretend that a raid had been undertaken when it had not. And there was no effective way to cooperate with the enemy in a raid because neither live soldiers nor dead bodies could be exchanged. The live-and-let-live system could not cope with the disruption . . . since raids

---

[3] Axelrod (1984, 71).
[4] Ashworth (1980).

could be ordered and monitored from headquarters, the magnitude of the retaliatory raid could also be controlled, preventing a dampening of the process. The battalions were forced to mount real attacks on the enemy, the retaliation was undampened, and the process echoed out of control.[5]

This is why, Axelrod argues, the truces broke down exactly then.

Is this a case, then, of Stability-Theorist paying off—does the Prisoner's Dilemma game indeed explain the truces? That is certainly how it is reported, including by Axelrod. But alas, the claim does not stand up.

To begin, by Axelrod's own admission, some elements of the story deviate from his predictions. The norms of most truces were not Tit-for-Tat but more like Three-Tits-for-Tat, that is, typically retaliation for the breach of a truce was roughly three times stronger than the original breach. (The Prisoner's Dilemma game itself—as opposed to the Tit-for-Tat strategy—is silent about the expected length of retaliation, so it should stand accused here merely of omission rather than error.)

More seriously, a crucial element to sustaining the truces was the development of what Axelrod terms ethics and rituals: local truce norms became ritualized, and their observance quickly acquired a moral tinge in the eyes of soldiers. This made truces much more robust and is crucial for explaining their persistence, as Axelrod concedes. Yet, as Axelrod also concedes, the Prisoner's Dilemma says nothing about it. Indeed, he comments (1984, 85) that this emergence of ethics is modeled most easily as a change in the players' payoffs, that is, *as a different game altogether*. I return to this below.

In addition to those remarked by Axelrod, there are several other important empirical shortfalls. First, his theory predicts there should be no truce breaches at all, but in fact breaches were common. Second, a series of dampening mechanisms therefore had to be developed to defuse post-breach cycles of retaliation, but the Tit-for-Tat analysis is silent about this crucial further element. Third, it is not just that truces had to be robust against continuous minor breaches; the bigger story is that often no truces arose at all. Ashworth examined regimental and other archives in detail to arrive at the estimate that, overall, there were truces only about one-quarter of the time—that is, on average, three-quarters of the front was *not* live-and-let-live.[6] The Prisoner's Dilemma is silent about why. Fourth, the Prisoner's Dilemma is also silent about how truces originated as opposed to how they persisted.

---

[5] Axelrod (1984, 82).
[6] Ashworth (1980, 171–5).

Overall, the Tit-for-Tat analysis is not empirically accurate; it misses crucial elements even in those areas where initially it does seem empirically accurate; and it is silent on obvious related explananda, some of them cited as targets by Axelrod himself: not just why truces persisted but also why they arose on only a minority of occasions, how they originated, and (to some degree—see below) when and why they broke down. On no philosophical account of explanation does the Prisoner's Dilemma explain here.[7] We have no empirical warrant that it identifies the relevant causes, which vitiates claims of causal explanation. And deductive-nomological, unification, and mathematical accounts similarly all require an empirical warrant that is absent here.

Axelrod is well aware that real-life complications make his analysis empirically inaccurate. He comments:

> The value of an analysis without [the real-life complications] is that it can help to clarify some of the subtle features . . . which might otherwise be lost in the maze of complexities of the highly particular circumstances in which choices must actually be made. It is the very complexity of reality which makes the analysis of an abstract interaction so helpful as an aid to understanding.[8]

Axelrod's meaning is a little ambiguous here, but I think his claims are best interpreted as some combination of partial explanation, heuristic value, and understanding, and maybe also the generalizability that any wide-scope explanation offers. Certainly, these are very reasonable goals. Indeed, if applying the Prisoner's Dilemma did not achieve any of them, what would be the gain from applying it at all?

How well does Axelrod's study fare by these criteria? Turn first to partial explanation, by which I mean an explanation that captures only one or a few explanatory factors and where those factors alone are not enough to explain all that we want to.[9] We need empirical warrant for any claim to have truly identified a cause and so to have achieved (partial) causal explanation. But such empirical warrant is just what is missing here.

If the incentives of the soldiers have the form of a Prisoner's Dilemma, then are Axelrod's explanatory claims vindicated anyway, never mind the problems above? No. First, this still would not explain why the truces hold in some cases and not others, nor how they break down, so at best only one portion

---

[7] Northcott and Alexandrova (2013).
[8] Axelrod (1984, 19).
[9] Northcott (2012, 2013b) analyze the relevant sense of partial explanation in detail.

of the overall truce phenomenon would be explained, contrary to advertising. Second, in any case, the Prisoner's Dilemma does not explain even this favored portion. We must distinguish between the strategic situation that faced the soldiers and the Prisoner's Dilemma's representation of that situation. What *was* the exact incentive structure in the trenches? The Prisoner's Dilemma is merely one suggestion—and upon closer inspection, a dubious one. Axelrod's arguments that the payoff structure fits an iterated Prisoner's Dilemma are rather brief and informal, and they have been doubted.[10] Further, Three-Tits-for-Tat is compatible with the strategic situation but not with Axelrod's analysis of it. And perhaps most tellingly, the ethics and ritualization needed to sustain the truces change the soldiers' incentives so that they no longer face a Prisoner's Dilemma at all—by Axelrod's own admission, recall. So, even if the strategic situation does partially explain the truces, that does not mean the Prisoner's Dilemma does.

More fundamentally, any explanation needs warrant, but that is what is lacking here. We do not have warrant to claim that the Prisoner's Dilemma is operating reliably but that its effects are concealed by a noisy environment. Why not? Because the relations captured by the Prisoner's Dilemma do not operate reliably. Meanwhile, there is no empirical success in the case at hand. So even if, in fact (but not established by us), the situation in the trenches *had* been a Prisoner's Dilemma, *we* would lack warrant for that claim—because of the lack of empirical success in the case at hand.

The Prisoner's Dilemma is wide scope. Models that are narrow scope, by contrast, can fit the evidence much more snugly. In the World War One example, the historian Ashworth achieves this by drawing on evidence specific to the case: regimental archives, letters sent by soldiers, and Ashworth's own interviews of veterans. The warrant for the causal explanations that result is narrow scope because the combination of circumstances that led to those explanations is easily disrupted, so they cannot be assumed to apply elsewhere. To establish that they do—for battlefields in other wars, for example—would require fresh investigation each time.

Ashworth explains in detail why the truces arose only in some parts of the front and not in others: the important distinction for that was between elite and nonelite units because these had different attitudes and incentives, for well-understood reasons, which meant truces usually occurred only between nonelite units. Next, why did breaches of truces occur frequently? Ashworth

---

[10] Axelrod's "rather brief and informal" arguments are at (1984, 75). Gowa (1986) and Gelman (2008) are two critics.

explains via detailed reference to different incentives for different units (e.g., artillery vs. frontline infantry) and to the fallibility of mechanisms for controlling individual hotheads.[11] Ashworth also explains how truces originated. Finally, he fills out how the system of truces came to an end. The escalating use of raids, so emphasized by Axelrod, is only one part of the story: many truces broke down even before this escalation. Ashworth devotes most of his Chapter 7 to a discussion of the intra-army dynamics, especially between frontline and other troops, which were the underlying cause of these breakdowns.

Ashworth also analyzes several examples of strategic sophistication that were important to the maintenance of truces but that are not mentioned by Axelrod and that go beyond any Prisoner's Dilemma analysis. One example is the use by infantry of gunners. Gunners were persuaded to shell opposing infantry in response to enemy shelling so that opposing infantry would then pressurize their own gunners to stop. This was a more effective tactic for reducing opponents' shelling than any direct attack on hard-to-reach opposing gunners.[12] Another example: the details of how increased tunneling beneath enemy trenches also disrupted truces, separately from increased raiding.[13] And even a full understanding of raiding goes beyond the simple Prisoner's Dilemma story. Ashworth summarizes: "Raiding . . . replaced a background expectancy of trust with one of mistrust, making problematic the communication of peace motives; raids could not be ritualized; the nature of raids precluded any basis for exchange among adversaries; and raiding mobilized aggression otherwise controlled by informal cliques."[14]

Removing our Prisoner's Dilemma lens, we see that we have perfectly adequate explanations already and ones that are confirmed much better. But if not explanation, perhaps the Prisoner's Dilemma offers other virtues? Return to the list from earlier. First, is the Prisoner's Dilemma, in Axelrod's words, an "aid to understanding?" It is much debated whether one may achieve understanding without explanation, but in any case, all agree that empirical success is needed and that is just what the Prisoner's Dilemma lacks here.[15]

Next, might the Prisoner's Dilemma instead earn its keep heuristically? Even if it does not itself explain, perhaps it guides us toward those strategic elements that do explain. But alas, the details of the case suggest not, for two reasons.

---

[11] Ashworth (1980, 153–71).
[12] Ashworth (1980, 168).
[13] Ashworth (1980, 199–202).
[14] Ashworth (1980, 198).
[15] See de Regt (2017) for a discussion of scientific understanding, and see also Chapter 11.

The first reason is that the Prisoner's Dilemma does not lead us to any explanations that we did not have already. To see this, note a dialectical jiu-jitsu. Axelrod gives many examples of soldiers' words and actions that seem to illustrate them thinking and acting in Prisoner's Dilemma–like patterns, to support the claim that the Prisoner's Dilemma explains. Yet now that we have rejected that claim and are concerned instead with heuristic value, these same examples become evidence *against* the Prisoner's Dilemma rather than for it. They show that soldiers were well aware already of the basic logic of reciprocity.[16] Soldiers were well aware, too, of why frequent raiding rendered truces impossible to sustain, an outcome indeed that many ruefully anticipated even before the policy was implemented.[17] Ashworth reports:

> One trench fighter wrote a short tale where special circumstances ... [enabled the truce system to survive raids]. The story starts with British and Germans living in peace, when the British high command wants a corpse or prisoners for identification and orders a raid. The British soldiers are dismayed and one visits the Germans taking a pet German dog, which had strayed into British trenches. He attempts to persuade a German to volunteer as a prisoner, offering money and dog in exchange. The Germans naturally refuse; but they appreciate the common predicament, and propose that if the British call off the raid, they could have the newly dead body of a German soldier, providing he would be given a decent burial. The exchange was concluded; the raid officially occurred; high command got the body; and all parties were satisfied. All this is fiction, however.[18]

This fictional tale demonstrates vividly a clear understanding of the strategic situation, indeed a more nuanced and detailed understanding than the Prisoner's Dilemma's own. No need for heuristic aid here.

The second reason to doubt the Prisoner's Dilemma's heuristic value is that it diverts attention *away* from the aspects that are important. I have in mind the crucial features mentioned above: how truces originated, the causes and management of the many small breaches of them, the importance of ethics and ritualization to their maintenance, why truces occurred only in some sections of the front, not others, and so on. Understanding these features is crucial if our aim is to encourage cooperation elsewhere—and this wider aim is the headline

---

[16] Ashworth (1980, 150).
[17] Ashworth (1980, 191–8).
[18] Ashworth (1980, 197).

one of Axelrod's book, and it is surely a major motivation for the Prisoner's Dilemma literature as a whole. Yet here, to repeat, the Prisoner's Dilemma actively directs our attention away from them!

Turn to the last virtue from earlier: generalizability. Perhaps the most common defense of wide-scope models is that, unlike narrow-scope explanations, they offer the virtue of systematization, as befits social science as opposed to history. Thus, the Prisoner's Dilemma, it is claimed, sheds light on cooperation in general, not just in the specific setting of the World War One trenches. But alas, this assessment is far too rosy. To be useful in many cases, it is necessary to be useful in individual ones.[19] Here, correctly understanding what encouraged cooperation in the World War One case is necessary for that example truly to teach us about cooperation in other cases too. But this is just what the Prisoner's Dilemma does not do.

It turns out that the Prisoner's Dilemma's failure in the World War One case is no fluke or outlier.[20] Trumpeted successes in biology dissolve under closer inspection: payoff structures in actual cases rarely match those of the Prisoner's Dilemma, often because of the different values put on an outcome by different players. Even advocates, such as the game theorists Martin Nowak and Karl Sigmund, concede that most claimed cases of the Prisoner's Dilemma occurring in nature are unproven: "it proved much easier to do [computer] simulations, and the empirical evidence lagged sadly behind."[21]

It is a similar story in psychology. Many experiments have tested variations of the Prisoner's Dilemma, but their focus has been on how, to predict cooperation, a formal Prisoner's Dilemma analysis needs to be *supplemented* by psychological and social factors. Besides, what we usually care about is not cooperation in psychology laboratories, but rather cooperation by banks, firms, consumers, and soldiers in the field. Are the laboratory experiments useful guides to that? So far, the evidence is not encouraging.[22]

This suggests skepticism about any generalization benefit. The underlying reason, as I discuss in Chapter 4, is that when relations are fragile, to gain purchase in actual cases, a model must be developed in the right way—roughly, by constant empirical refinement. But the Prisoner's Dilemma has not been developed in this way, as I will discuss shortly. That is why it is the wrong tool for the job.

---

[19] Northcott (2017).
[20] See Northcott and Alexandrova (2015) for more detailed discussion, and references, regarding this and the next paragraph.
[21] Nowak and Sigmund (1999, 367).
[22] Levitt and List (2007); Diener et al. (2022).

Axelrod's analysis of the iterated Prisoner's Dilemma via his computer tournaments has been criticized by other game theorists. They dispute whether Tit-for-Tat is the uniquely favored strategy or even a favored strategy at all.[23] But no successor game-theoretical analysis of the World War One truces has appeared, and why believe that, if one did, it would explain the behavior in the trenches any better than Axelrod does, let alone better than Ashworth does?

Might the Prisoner's Dilemma still be useful elsewhere? For example, from the start, it was lauded for illuminating how individual and social optimality may diverge. But there is reason to be cautious. As the World War One case illustrates, and as seems to be a pattern generally, it is all too easy casually to claim heuristic or explanatory value for the Prisoner's Dilemma when in fact its value is zero or even negative.

Another defense of the Prisoner's Dilemma, different from Axelrod's, is that it plays a *normative* role. With a normative role, the goal is no longer an accurate description. For example, even when, in fact, soldiers did not cooperate, perhaps this shows merely that they acted unwisely, not that the Prisoner's Dilemma's analysis is vitiated. Similar defenses have been made of game-theoretical and rational-choice models generally.

In reply, I am all in favor of normative analysis, but it should not be abused to excuse bad science. No one disputes that, given their desire to live and given the situation they faced in the trenches, the soldiers should have created truces if they could. But the soldiers themselves were well aware of that. It is insulting to them to imagine the Prisoner's Dilemma was needed to teach them that cooperation was desirable. Rather, the pressing issue was not the desirability of cooperation but how to achieve it, and for that, the Prisoner's Dilemma's crude analysis does not help, because it misses the crucial details that made cooperation possible. Some degree of empirical accuracy is required even for normative purposes.

Another way to make the point: the Prisoner's Dilemma's normative analysis is compelling only if the target situation instantiates a Prisoner's Dilemma. Still, it might be objected that the analysis at least shows that the soldiers would have been well advised to change the situation *into* a Prisoner's Dilemma. But even this follows, only if changing the situation into a Prisoner's Dilemma was indeed the best way to achieve cooperation, which in turn depends on the empirical details—and those details suggest the advice is fanciful. Much the best routes to cooperation were those the soldiers actually pursued: ethics and ritualization, mechanisms to control individual hotheads, manipulation of

---

[23] Again, see Northcott and Alexandrova (2015) for discussion and references.

intra-army dynamics, and so on. The underlying point, again: there is no normative escape route from the need for empirical accuracy.

There is one further, important issue that the World War One case study illustrates: use of resources. There has arisen a gross disproportion between a huge Prisoner's Dilemma literature and the meager number of explanations of field phenomena that this literature has yielded. According to JSTOR, at the time of writing, over 29,000 journal articles about the Prisoner's Dilemma have appeared since 1970, in addition to over 14,000 book chapters and 500 research reports. A striking aspect of this literature is its overwhelmingly theoretical focus. Much of it concerns mathematical developments of the basic game: asymmetric versions, versions with multiple moves or players, single-person interpretations, versions with asynchronous moves, finitely and infinitely and indefinitely iterated versions, iterated versions with error, evolutionary versions, versions interpreted spatially, and many other tweaks besides. The subtlety and sophistication of this work are impressive, even bewitching, and it feels philistine to criticize. But it is philistine *not* to. For to endorse this allocation of effort is, in effect, to turn our faces against other lines of research that work better.

Empirical applications of the Prisoner's Dilemma are notably thin on the ground. In fact, it is hard to find serious attempts to apply it to explain actual historical or contemporary phenomena, as opposed to informal mentions or off-hand remarks.[24] Axelrod's analysis of the World War One truces is one of the few exceptions, which might account for its fame and frequent mention in textbooks.

The pattern of theory development in the Prisoner's Dilemma literature—driven by internal logic rather than empirical applications—is a paradigm of how, I argue in Chapter 4, theory should *not* be developed when relations are fragile. Without continuous empirical application, such theory development generates a poor harvest of explanations, predictions, and interventions. It is inefficient. In this book's terminology, theory should instead be developed using a Case-Worker strategy, not a Stability-Theorist one.

The World War One case study illustrates, finally, how "one more heave" can be a harmful pipedream. The Prisoner's Dilemma's shortcomings are due to a mismatch between a theory with fragile relations and a method of theory development suited to stable ones. The result is a square peg in a round hole. Greater contextual detail is the only remedy, a task for which "one more heave"

---

[24] Sunstein (2007) comes close, but even here the target phenomenon (the failure of the Kyoto protocol) is explained in part by the fact that it does *not* have a Prisoner's Dilemma structure.

of ever more theoretical development is conspicuously unhelpful—good money after bad. If our currency is explanations of real-world phenomena, resources directed to the Prisoner's Dilemma are not well spent. They are better directed to the history department.

3.2 Underdetermination in field sciences, and how to overcome it

My focus in this chapter is field sciences, by which I mean nonlaboratory investigations of systems that are not engineered artifacts.[25] I will discuss situations in which the explanans relation is fragile. I concentrate on causal explanations—not because I rule out other kinds of explanation but because, in most cases we will be looking at, the models and explanations of interest are causal.

The most common template for causal inference is, roughly, to vary one thing while keeping others fixed. What we might call *scientific* prediction concerns the results of such manipulations (or when the right conditions arise without manipulation, as in natural experiments).[26] So, not all predictions are scientific. A prediction is in addition scientific only if it concerns data gathered in certain epistemologically propitious circumstances. Scientific predictions may be conditional—about what would happen if we intervened in a certain way, for example. As I will use the term, a prediction may be retrospective as well as forward-looking.

Field environments are typically noisy. By "noisy," I mean it is hard if not impossible to model all of the significant factors in play. In turn, this is because situations cannot easily be shielded, meaning both that there are likely many disturbing factors and that we cannot control which ones affect our target. As a result, it is hard to isolate the impact of one factor alone, so a ubiquitous problem in field sciences is to distinguish signal from noise. A model might posit that X causes Y, so to test the model, we must measure the impact of X on Y. But if other, unmodeled factors A, B, and C ("noise") also impact Y, then scientific prediction is not simply predicting Y after a change in X, because we need to shield off, or control for, the influence of A, B, and C too. In field settings, such shielding off is difficult, and therefore so is scientific prediction.

---

[25] Parts of the next few sections are adapted with permission from Northcott (2023).
[26] I take the term "scientific prediction" from Dowding and Miller (2019). See also Popper (1989); Salmon (1981); and Watkins (1968).

The problem of noise is familiar, of course. It has familiar solutions: natural experiments, quasi-experiments, randomized field trials, and laboratory experiments.[27] All of these achieve scientific prediction, by exploiting fortuitous circumstances (natural experiments), balancing the impact on treatment and control groups of unmodeled factors (quasi-experiments, randomized field trials), or shielding the target (laboratory experiments). For when these solutions are impossible or unavailable, an array of other statistical techniques has been developed to infer causes from nonexperimental data. These various solutions and techniques each have their own strengths and weaknesses, but when relations are fragile, two serious problems afflict them all. The first problem is limited scope: experiments can elucidate only some field science questions, for practical and ethical reasons, while as I discuss in Chapter 6, statistical methods are difficult to apply to explanatory claims that are narrow scope. The second problem is external validity: if a relation is fragile, then even if we infer that it is operating in one case, that inference does not extrapolate reliably.

These two problems make inference difficult. That breeds *underdetermination*. For example, why did Joe Biden defeat Donald Trump in the 2020 US presidential election? Some of the many suggested explanations: the COVID-19 pandemic, Biden's appeal to Midwest swing voters, Trump's unpopularity, voters' economic insecurity, hostile media coverage of Trump, and anti-Trump mobilization in swing states. The mere headline fact that Biden won does not by itself discriminate between these explanations, because—by design, of course—they are all compatible with it. No experiment, or statistical study of many Biden–Trump elections, is available. Underdetermination is especially acute if the explanandum is qualitative: why did Biden win, rather than why did he win by a certain margin? But even in quantitative cases, the bar is lowered when only a partial or approximate fit with the data is demanded. The Biden example is not an outlier: underdetermination is frequent.[28]

Logically speaking, indefinitely many models fit any given body of evidence. But what matters practically is whether, as in the Biden case, these different models are also plausible or to be taken seriously.

If experiments and statistical studies are unavailable or inapplicable, how else might we solve the underdetermination problem? One proposal is forward-looking prediction: if you stick your neck out in advance, the proposal argues, that rules out fudging awkward outcomes after the fact. For example, many

---

[27] Northcott (2019).
[28] Dowding (2016) endorses this claim for the case of political science.

models or pundits might explain retrospectively the outcome of one election, or even of five elections, relatively plausibly, but few can predict correctly five in advance.[29] There is a huge literature on different versions of predictivism—roughly, the view that privileges forward-looking prediction over retrospective accommodation. I do not discuss that literature here, but in summary: opinion is split.[30] Perhaps sometimes, in some circumstances, predictivism does solve the underdetermination problem, but even when it does (if it does), if the relevant relation is fragile, then accurate forward-looking prediction requires extensive contextual knowledge. And the supplementary investigation needed to acquire such knowledge, finally, is in any case the other solution to underdetermination that I will promote now. So, whether we are predictivists or not, supplementary investigation is the way to go.

At root, solutions to the problem of underdetermination are all ways to establish that there are *no plausible alternatives* to a favored answer. If predictivism is right, this is why—only one model or theory predicts successfully, even if many can accommodate retrospectively. But accurate prediction is hard to achieve in field cases, so even if we accepted predictivism, it would not fully solve underdetermination.

If prediction is not the solution, what is? The answer: to gather *additional evidence* that favors one explanation over others. This is a version of the no-plausible-alternatives solution, but now with respect to an augmented body of evidence. Scientific prediction is enabled. In the case of an election, for example, the additional evidence might be post-election interviews of voters or comparisons of vote shares in different districts cross-referenced by potentially relevant demographic and economic variables. In this way, we break the epistemic tie.

Happily, we usually can gather such additional evidence. Historians do it all the time. To establish what explained the World War One truces, the historian Ashworth read archives and interviewed veterans. If the relation in our putative explanation is fragile, this supplementary evidence must be specific to the case. That is why Ashworth interviewed only veterans of the World War One trenches. Warrant for an explanation is then also specific to the case—in other words, the warrant is *narrow scope*. (More on this in the next section.)

---

[29] I will speak of predictive or empirical 'success' and 'accuracy'. These are a matter of degree, and the appropriate standards for assessing them are contextual, but the judgments here will be clear enough for our purposes.

[30] Syrjänen (2022) is an excellent survey and discussion.

Such localism brings with it a challenge. Formal techniques of causal inference, such as experiments or statistical analysis, are usually inapplicable, so is localist causal inference enough to overcome underdetermination? In reply, well-established procedures of qualitative causal inference may be applied to single cases.[31] As I discuss in Chapter 6, it is not clear that these procedures are less rigorous than quantitative ones, and it is certainly dubious to claim that they always are. There are also well-established methods of measurement in field sciences, developed over many decades.[32] And historical sciences have their own rich epistemology, which contributes to widespread consensuses about matters of historical fact.[33]

Some historical controversies do persist, nevertheless. These sometimes turn on concerns beyond just fidelity to the evidence, of course, but our focus here is only on controversies about the latter. If, even after contextual historical work, matters are still empirically underdetermined, then we are stuck. We must concede we just do not know which explanation is correct.

Total skepticism about narrow-scope causal inference would negate any prospect of narrow-scope causal explanation. What is at stake, therefore, is whether science can succeed at all when relations are fragile, because if we deny that narrow-scope causal inference is ever justified, then it cannot. It is fortunate that such extreme skepticism is implausible.

3.3 Fragility, warrant, and narrow scope

We have seen one reason why, when we explain with fragile relations, warrant is narrow scope: to overcome underdetermination, investigation needs to be contextual, and explanations become specific to a particular case. But there are other reasons, too.

Begin with an example from physics: Coulomb's Law. This entails that two oppositely charged bodies attract each other. Even if, out in the field, disturbance by other forces means that the two bodies do not move toward each other, still we are confident that the electrostatic attraction operates. Why? Because Coulomb's Law is vindicated by laboratory experiments and because of a stability assumption that the electrostatic force observed in the laboratory

---

[31] Brady and Collier (2010); see also Rubin (2021).
[32] Boumans (2015).
[33] Tucker (2004); Chapman and Wylie (2016); Currie (2019a).

continues to operate in the field even when other forces are present too.[34] These give warrant that, even when Coulomb's Law does not predict accurately the motions of charged bodies outside the laboratory, it still explains those motions partially, in the sense that it captures *one* of the forces at work.

This strategy of importing warrant from other contexts allows us to explain even without empirical success here and now. Typically, field situations are unshielded, so we cannot control the range of causes present, and so empirical success here and now requires accounting for many uncontrolled causes. That makes empirical success here and now hard to come by; it is a huge advantage not to need it.

But importing warrant does not work if a relation is fragile. The reason is epistemological: if a relation is fragile, then we cannot infer that it operates here just because it operates elsewhere. Just because voter worries about immigration influenced one election, for example, that does not establish that they influenced a different election too (if this influencing relation is fragile). And just because agents behave in accordance with the Prisoner's Dilemma in an experiment, that does not mean that soldiers will when at war. Success in the laboratory is not enough. Unlike with Coulomb's Law, we always need fresh empirical success here and now; warrant cannot be had any other way. It follows that warrant is narrow scope.

All this is true even if the relations hold reliably *in the world of the model* because that does not imply that those relations hold reliably in the actual world, and certainly not that we have warrant for this latter claim. The form of a model can mislead.

Importing warrant is threatened in a second way, too. In noisy field environments, many times there is no empirical success *any*where, and thus no empirical warrant from elsewhere is available to *be* imported. There is no analogue of the successful Coulomb laboratory experiment.

Here is another way of seeing why, when relations are fragile, warrant is narrow scope. In field sciences, environments are unshielded, which means that many and varied causal factors can be at play, each context an idiosyncratic smorgasbord of them. Wide-scope models inevitably miss sui generis local factors, and so they identify only some of the causal relations present in each field environment. Empirical vindication therefore eludes them. Only explanations tailored to the specific circumstances of a case, such as Ashworth's of the World War One truces, stand any chance of such vindication and, thus, any chance of gaining warrant.

---

[34] Cartwright (1989).

This localist picture dovetails with recent theories of causal explanation, such as James Woodward's, which are framed not in terms of general laws but rather in terms of invariance relations that may be of highly restricted scope.[35] As we will see, it also dovetails with much other recent philosophy of science, which has emphasized the need for local work to know when and how models apply.

Perhaps, it might be objected, wide-scope models can still explain partially, because they capture *some* of the causes present. Indeed, they can. And there is no shame in that: in noisy environments, such partial explanations are often the best we can hope for. But a partial explanation requires warrant just the same as any other explanation, and if an explanation is empirically inaccurate in the case at hand, this warrant must be imported from empirical success elsewhere, just as applications of Coulomb's Law outside the laboratory import warrant from experiments inside the laboratory. And as we just saw, fragility undermines this "imported warrant" strategy, which leaves no warrant for wide-scope explanations even when they are partial. That is why the Prisoner's Dilemma's claims to explain the World War One truces were without warrant. Only narrow scope is left.

For example, suppose we explain Biden's 2020 election victory by saying that the incumbent, Trump, was hurt by low gross domestic product (GDP) growth in the preceding year. Many other factors too are causally relevant, so the GDP explanation will at best be only partial. If we could somehow tweak the United States in 2020 so that only GDP was altered and run the election again, then we could test the GDP explanation and *only* that explanation. Obviously, such an experiment is impossible. In principle, evidence collected across many elections could help, but that would have to assume that any impact of GDP operates stably across elections, which is dubious. In which case, importing warrant from other elections is ruled out. And anyway, similar epistemic barriers arise in other elections too, so there is no warrant from elsewhere to be imported. The same problems afflict other candidate explanations of Biden's win too. How demographic factors impact voter preference, for example, changes from election to election, and even during elections and between different regions.[36] The only solution is empirical accuracy in the case at hand, which implies warrant that is narrow scope.

---

[35] Woodward (2003).
[36] Northcott (2020).

In sum, given noise and *stability*, warrant for explanations, or at least for partial explanations, can still be wide scope, after the manner of Coulomb's Law. But given noise and *fragility*, it cannot.

3.4   Fragility and prediction of the future

In J. K. Galbraith's cynical one-liner, "economic forecasting is there to make astrology look good." Why is accurate prediction in field sciences difficult? (In this section, unlike elsewhere in the book, by "prediction" I will mean specifically prediction of the future.)

First, target systems are often complex—many factors are in play, they interact in many and changing ways, and many are difficult to observe. If a target system is chaotic, in the technical sense that outcomes are extremely sensitive to initial conditions, then, for humans, at best only probabilistic forecasts are possible, even in principle. Second, target systems are unshielded. The economy, for example, is continuously impacted by noneconomic factors, such as election results, which inevitably do not appear in economic forecasting models—as long ago as 1928, Oskar Morgenstern pointed out that economic prediction requires prediction also of noneconomic variables. Third, in the case of economic targets, market prices might already incorporate all public information, so changes in them are impossible to predict. This is why many rational expectations models in economics deem it impossible to forecast systematically better than a random baseline. Fourth, data—the raw ingredients of any forecasting model—are often bad. GDP, for example, can be estimated only by aggregating meso-level inputs, which requires many statistical estimates and subjective judgments. Methods for seasonal adjustment introduce further imprecision. One symptom of these difficulties is significant discrepancies between different measurement methods. Another symptom is the large size of revisions, which for GDP are typically greater than 1%—comparable to the average forecast error.[37]

What do these reasons have in common? Fragility. The first three reasons are *causes* of it. First, if the ways in which factors interact are complex and changeable, then predicting when relations between these factors will hold is more difficult. Second, if a relation faces disruption by unmodeled external factors, then whether it holds will be harder to predict. Third, if the causes of a price movement cannot be captured by any model using public information, then

---

[37] Betz (2006, 101–8).

their operation will again be harder to predict. With all three of these, overall prediction is made difficult because it is difficult to predict when the relevant relations will hold. The fourth reason, meanwhile, namely bad data quality, is typically an *effect* of fragility. When calculating GDP, for example, the relations behind the correct calibration of seasonal adjustments operate unpredictably, hence the frequent errors and need for later revisions.

Another reason commonly cited for the difficulty, or even impossibility, of prediction, at least in social science, is *reflexivity*: when the act of prediction itself impacts the target system. Any system featuring human agents is potentially subject to such effects. But I think that to emphasize reflexivity is a mistake because fragility matters much more.[38] I save a full discussion of that, though, until Chapter 7.

Are these difficulties a counsel of despair? Not always. At the macro level, true enough, the record mandates caution. In political science, for example, predictions of civil wars and election results have proved disappointing,[39] while in macroeconomics, predictions of GDP, inflation, and exchange rates are no better.[40] The record is better at the micro level, though. Why? Because a route to accurate predictions is local knowledge. Combined with relevant background knowledge, local knowledge warrants forward-looking predictions when it can establish that most of the significant causal relations have been modeled and that they are stable enough.[41] This localist method obviously favors the micro level. For example, local knowledge might warrant a confident prediction of which political party will win an election in a new district, even when at the national level the overall result remains in doubt.

There are general reasons, too, to expect accurate prediction to require local knowledge. First, prediction often requires causal knowledge, and if relations are fragile, then we should expect warrant for causal knowledge to be narrow scope, for the same reasons that warrant for causal explanations is. The literature on extrapolating a model from one context to another, including from the past to the future, supports this: such extrapolation requires detailed knowledge of the target context.[42]

More basically, such localism is just what works. A lesson of Philip Tetlock's well-known forecasting tournaments is that prediction is best done contextually. Winning forecasters open-mindedly incorporate local or case-specific

[38] Northcott (2022b).
[39] Tetlock (2005); Ward et al. (2010); Northcott (2015).
[40] Betz (2006); Meese and Rogoff (1983).
[41] Sterelny (2016).
[42] Cartwright (2019); Khosrowi (2019b); Cartwright and Hardie (2012).

knowledge; simply ploughing ahead and applying one's favorite wide-scope model or theory regardless, by contrast, is a losing strategy.[43] Localism is also the method of successful sports gamblers and other predictors.[44]

A localist moral is the lesson of prediction markets, too. Prices in prediction markets are implicit probabilistic predictions, and such markets have a sustained record of relative success in many domains.[45] How? In effect, prices are set by an informed minority of traders who integrate multiple lines of evidence case by case, just as Tetlock's forecasting tournament winners do. The two takeaways, as for field predictions generally: first, some predictive success is achievable; and second, when it is, it is achieved by localism.

In sum, despite noise, there is hope—but only via empirical success, which means (when relations are fragile) only via localist work, which in turn means that warrant for explanations and predictions is narrow scope. It might be objected that even single cases need wide-scope theories for their analysis. Indeed, they do. In the next chapter—Chapter 4—I discuss in detail the role for theory that results. The lesson will be that theory must be both applied and developed in the right way—a way that accounts for fragility.

[43] Tetlock and Gardner (2015).
[44] Silver (2012).
[45] Tziralis and Tatsiopoulos (2007); Horn et al. (2014).

4

The core divide: Stability-Theorist versus Case-Worker

4.1 Terminology

This chapter forms the heart of the book. In it, I develop an analysis of the core methodological divide behind the book's approach and of the implications of that divide. At the end of the chapter, I connect the book's analysis to its many forebears in recent philosophy of science.

Begin by clarifying terminology. I use "theory" as an umbrella term for descriptions of the world that are wide scope. There is a methodologically important distinction between high-level theories and middle-level theories, where the latter refers to entities such as mechanisms, ceteris paribus laws, and nomological machines.[1] I return to this distinction later in the chapter. I use "theory" to cover both kinds of theory alike.

On some views, theories are best understood not just as sets of propositions or models but rather as comprising also practices, equipment, measures, know-how, and other less formal features.[2] I sympathize with the picture of science behind these views, as will become clear. But for clarity, I stick here with a narrower understanding of theory.

In many sciences, *models* are central. The study of them has flourished in recent philosophy of science: what kinds of model are there, how do we learn and explain with them, how do they represent targets more complicated than themselves, and what are they ontologically?[3] Much of the book is written on the back of this valuable work. Although the exact relation between models and theories is debated, all agree that the two are importantly distinct. Nevertheless, I do not focus on that distinction because I think the more significant distinction methodologically is that between fragility and stability. The two distinctions do not map onto each other neatly. Finally, even though

---

[1] Cartwright (2020a).
[2] Cartwright (2020b); Cartwright et al. (2022); Leonelli (2016).
[3] Morgan and Morrison (1999); Weisberg (2013); Frigg and Hartmann (2020).

Science for a Fragile World. Robert Northcott, Oxford University Press. © Robert Northcott 2025.
DOI: 10.1093/9780191944352.003.0004

models are at the heart of many of my examples, for ease of exposition, I often use "theory" to refer to models.

4.2 Theory: a toolbox role

Theory is essential to science. It is a truism of philosophy of science, and of philosophy generally, that analysis even of one-off cases must borrow from background knowledge, that is, from theories.[4] How may we reconcile this with the localism of Chapter 3? The answer is *warrant*: the warrant for an explanation derived from a theory may be only narrow scope even if the theory itself is wide scope.

Consider Ashworth's historical explanations of the World War One truces. The warrant for these explanations is narrow scope, as they are established only for the World War One case, but they appeal to psychological and other theories that are wide scope. According to Ashworth, for example, breaches of truces were frequent in part because the mechanisms for controlling individual hotheads were fallible. This explanation rests on many wide-scope claims, such as that some humans ("individual hotheads") are prone to behave disruptively, that these humans will moderate their disruptive behavior in the face of disapproval by local superiors, but that this moderating does not always work ("mechanisms were fallible"). Ashworth established that in the World War One trenches the moderating indeed failed sometimes, and that when it failed, the resultant hotheaded actions could cause a truce to break down.

Wide-scope theories are essential to this explanation. Why, then, is the explanation's warrant only narrow scope? Because we cannot assume in advance that the mechanisms on which it rests, will hold. Why not? Because those mechanisms are fragile—they hold unpredictably. That is why Ashworth had to interview veterans, read archives, and so on, to discover when the mechanisms operated and when they did not, or when and how they were overridden. And that is why the warrant for the explanation is narrow scope, even though the mechanisms are not. This localist conclusion is reinforced by the details of how explanations are established, which often go well beyond just a simple judgment of whether a theory applies or not, as we will see in the next section.

So, the contribution of wide-scope theories to narrow-scope explanations can be accommodated. But there is a price. That price is, roughly, that we no

---

[4] E.g., in their different ways, Cartwright et al. (2022); Woodward (2003); Davidson (1963); Hempel and Oppenheim (1948).

longer care whether theories are true. Less roughly, we care only whether a theory is true *of the case at hand*; its truth elsewhere is irrelevant. We should think of a *toolbox*.[5] The more theories in the toolbox, the more resources with which to construct an explanation in a specific case. The more psychological mechanisms available to Ashworth, for example, the more resources he had to explain the World War One truces.

On the toolbox view, theories lose exclusivity. Many different theories may be useful in the same domain—no single master theory, Newton-style; rather, fragmentation.

The toolbox view of theory has several philosophical forebears. A version of it was put forward almost 30 years ago by Nancy Cartwright, Towfic Shomar, and Mauricio Suarez.[6] Another forebear is the "context-mechanism-outcome" frame of evaluation defended by Ray Pawson and Nick Tilley, which grows out of the critical realism of Roy Bhaskar.[7] Marcel Boumans similarly emphasizes extra-theoretical elements unrelated to a parent theory, such as mathematical formalisms, empirical data, metaphors, and analogies.[8] More recently, Nancy Cartwright and collaborators have urged that theory is but one part of a wider "tangle," comprising also many other elements.[9]

How does one evaluate a theory if failure in an individual case is no longer fatal? By how well it plays its toolbox role. We must assess whether a theory earns its keep overall, across many cases, relative to alternatives and predecessors—how many successful predictions, interventions, and explanations does it enable? How useful is it, how easy to use, how adaptable, and how deep? The verdict is not always clear-cut immediately, but usually it becomes so eventually.

4.3 Example of theory application: economic auctions

Not only does a theory's truth matter only in the context of interest; sometimes, it does not matter even there. The truth of an eventual *explanation* certainly does matter. So too, does the accuracy of a prediction or intervention. But sometimes, explanations, predictions, and interventions are not derived

---

[5] I take the term "toolbox" from Cartwright et al. (1995). Others have used the term, too (Ylikoski 2019).
[6] Cartwright et al. (1995).
[7] Pawson and Tilley (1997); Bhaskar (1975).
[8] Boumans (2015).
[9] Cartwright (2020a, 2020b); Cartwright et al. (2022).

directly from any parent theory; rather, they emerge from a mélange of many ingredients, of which a parent theory is just one. An example illustrates.[10]

The radio spectrum is the portion of electromagnetic spectrum between 9 kilohertz and 300 gigahertz. In the United States, parts of the radio spectrum that are not needed for governmental purposes are distributed via licenses by the Federal Communications Commission (FCC). For a long time, most of these licenses were awarded on the basis either of politicized hearings in which potential users—usually telecommunications companies—had to demonstrate the public interest of their proposed enterprise or else simply by lottery without any regard for economic suitability. But in the early 1990s, the FCC gained the right to use instead competitive market mechanisms such as auctions.

That left the FCC the formidable task of designing and running such auctions. The importance of this is best illustrated by what happens when it goes wrong. Notorious examples include when an Otago university student won the license for a small-town TV station by bidding just $5 (New Zealand 1990); when an unknown outbid everyone but then turned out to have no money, thus delaying paid television for nearly a year for do-over auctions (Australia 1993); and when collusion and a subsequent legal fight resulted in four big companies buying the four available licenses for prices only one-fifteenth of what the government had expected (Switzerland 2000). By contrast, the FCC's series of seven auctions from 1994 to 1996 were a remarkable success. They attracted many bidders, allocated nearly 2,000 licenses, and raised an amount of money—$20 billion—that surpassed all government and industry expectations. Even the first auctions passed off without a glitch, and there was reason to believe that licenses were allocated efficiently.

What matters for our purposes is how this successful design was arrived at. A wide range of goals was set by the government, beyond just maximizing revenue. These included efficient and intensive use of the spectrum, promotion of new technologies, and ensuring that some licenses went to favored bidders such as minority- and women-owned companies. Exactly what design would achieve these outcomes was an intricate puzzle for teams of economic theorists, experimentalists, lawyers, and software engineers. The country was divided into 492 basic trading areas, each of which had four spectrum blocks up for license. The eventual design auctioned all of these licenses simultaneously as opposed to sequentially, in an open rather than sealed-bid arrangement. Bidders placed bids on individual licenses as opposed to packages of licenses.

---

[10] This section is adapted with permission from parts of Alexandrova and Northcott (2009). See also Guala (2005) and Alexandrova (2008).

When a round was over, they saw what other bids had been placed and were free to change their own combinations of bids. Bidders were also constrained by further rules, such as upfront payments, maintaining a certain level of activity, increasing the values of their bids from round to round by prescribed amounts, and caps on the amount of spectrum that could be owned in a single geographical area. The full statement of the auction rules ran to over 130 pages.

At the time, this gleaming success was hailed by the press as a triumph for game theory, which had revolutionized the auction literature in the 1980s. And indeed, many game theorists were hired by prospective bidders or by the FCC itself. The final spectrum auction design, though, was not derivable from game theory alone. No single model covered anywhere near all of the theoretical issues mentioned above. In addition to those issues, explicit and public instructions were required to cover entry, bidding, and payment, and much work was also put into other features such as the software, the venue and timing of the auction, and whatever aspects of the legal and economic environment the designers could control. Because a particular design's efficacy could not be predicted from auction theory alone, many experiments and consequent fine-tuning were essential. The results of these experiments often took designers by surprise. For example, in some circumstances—against theoretical predictions—"bubbles" emerged in the values of the bids. That is, under the competitive pressure of an auction, prices of spectrum blocks were bid up well beyond their apparent economic value, much like with stock market bubbles. These bubbles in turn were unexpectedly sensitive to what information about rival bidders' behavior was available. The experiments revealed something that could not have been known just from theory. Chief experimental investigator Charles Plott commented:

> Even if the information is not officially available as part of the organized auction, the procedures may be such that it can be inferred. For example, if all bidders are in the same room, and if exit from the auction is accompanied by a click of a key or a blink of a screen, or any number of other subtle sources of information, such bubbles might exist even when efforts are made to prevent them. The discovery of such phenomena underscores the need to study the operational details of auctions.[11]

Experiments showed the ubiquity of interaction effects. That is, the impact of an auction rule varied, depending both on how it was implemented and on

[11] Plott (1997, 620).

which other rules were included. Because individual rules did not have stable effects across different environments, the performance of any *set* of rules had to be tested holistically, and retested with every significant change in environment. The eventual result was the perfecting of one overall design package, that is, of a set of formal rules and practical procedures together. It was not a matter of stitching together component capacities or mechanisms.

The auction design was *not* a case of tinkering with various models' parameters or of making a specific model work via painstaking trial and error. The designers had to go well beyond that. What was eventually made to work was instead a messy composite of many separate models plus much else besides. The same is true of other successful spectrum auctions, such as the 2000/1 ones in the UK.

The auction's success gives no empirical warrant for any of the game-theoretical auction models, rather only for the final auction design—which, to repeat, was entirely distinct from any model or combination of models. Any realist inference from empirical success to truth applies only to this final design. Thus, the success of the auction may warrant realism about, say, the causal claim that the US design led to greater revenues than did the Swiss one, but not realism about the relevant economic theory. After all, the designers of the US and Swiss auctions had access to essentially the same theoretical repertoire. The difference lay elsewhere.

Theory's role here is at best heuristic. For the experimentalist Plott, the game theory models were useful for generating categories with which to start, but the design process then took on a life of its own with the experimental testing and refining. Label this use of theory *heuristicist*.[12] The contrast to heuristicism I will label (in the case of causal theories) *causalism*—when a theory truly identifies some causal relation, and so directly explains a real-world target. This is the case familiar from textbooks, as when a Newtonian model directly identifies an explanatory force, such as the gravity that causes an object to fall. With the spectrum auction, such direct explanation is just what the auction models did *not* offer. Heuristicism and causalism are two distinct routes by which a theory may achieve the same goal, namely, to facilitate explanations, predictions, and interventions. A heuristicist role exemplifies the toolbox view of theories.

In the spectrum auction example, heuristicism is the only account of how theory is used that matches the details of the case.[13] Consider alternative

---

[12] Alexandrova (2008) gives a formal account.
[13] See again Alexandrova and Northcott (2009) for detailed discussion.

accounts. The relevant theory here is the game-theoretical auction models, but, first, there was no partial isomorphism between these models and the successful auction design.[14] Second, nor was the design achieved by de-idealizing the models—the models' assumptions never were satisfied.[15] Finally, third, nor did the auction models provide stable capacities or tendencies about the effects of auction rules.[16] No such stability existed. This is why the experiments were needed—interaction effects between different rules could be overcome only by establishing a material environment in which the auction worked as desired. Material environments had to be treated as *wholes*, in which all elements together produce the result.

For example, in some models, an open auction design (i.e., no sealed bids) reduces the "winner's curse" effect, by which the winner of an auction overpays because they are the ones who most overestimated the prize's true value. But would this effect of open auctions still hold when the value of a spectrum license depends on what other licenses a bidder wins? No one knew. At the time, no model that incorporated such interdependence even existed. In the experiments, more features were added specifically to defeat the winner's curse, such as flexible minimum increments that made it more attractive to bid when activity was low and activity rules that required bidders to submit bids on pain of losing their eligibility. The point: designers could not trust that the capacity of an open auction to defeat the winner's curse was stable enough to be relied on.

Not all cases are heuristicist cases. But in such cases, theory's toolbox role is especially apparent.

Whether the use of theory is heuristicist or causalist, either way, fragile explanans relations mean that warrant for explanations is narrow scope. First, the less idealized and the more contextual a model, the more likely it is we can read actual causes from it directly; in other words, the more likely it is we can apply it in a causalist way. But the more contextual a model, the more likely that an explanation derived from it is narrow scope. Second, the warrant for explanations achieved via the heuristicist route is usually narrow scope too, because such explanations rely on contextual investigation over and above theory. For example, because the predictions and explanations that the US spectrum auction design provided were narrow scope, they did not automatically carry over to new cases, and the 2000 spectrum auction in Switzerland—which occurred

---

[14] For a partial isomorphism account, see DaCosta and French (2003).
[15] For a satisfaction-of-assumptions account, see Hausman (1992); Morgan (2002).
[16] For a capacities or causal tendency account, see Mill (1843); Cartwright (1989); and Mäki (1992).

after the US spectrum auction and indeed was inspired by that auction's success—could be a failure.

4.4 The core divide: two ways to apply theory

How can theory best play a toolbox role? Two core errors must be avoided: (1) claims of explanation and understanding that lack warrant and (2) theory developed unfruitfully because developed abstractly, that is, without continuous empirical refinement. Both errors, alas, are common. To see how to avoid them, I set out now the methodological dichotomy at the heart of this book, because both errors have the same cause: when the relations in a theory are fragile, but the Stability-Theorist rather than Case-Worker strategy is followed.[17]

4.4.1 Stability-Theorist strategy

Begin with a simplified, benchmark case of a relation that holds reliably. Suppose that, to predict the motion of a moon, we apply a Newtonian two-body model of gravity, inputting the moon and parent planet's masses, positions, and motions. (By saying a model or theory "applies," I will mean that a model correctly, or—as with Newtonian theory—approximately correctly, represents a force or cause in the target system. So, when a model applies, it explains, at least partially, and guides interventions.) Something like this procedure is a staple of actual space exploration, and it works startlingly well. Why? Because of predictability: the Newtonian model that has been successful elsewhere may safely be assumed to apply to the new case, in turn because gravity itself may safely be assumed to be operating in the same way. Just reapply the same master model. The power of this strategy is greatly enhanced because the same model also successfully handles (gravitational) disturbances: posit another body and then use the model to calculate the impact of that body.[18]

Because this strategy assumes that the relations in a theory or model hold stably, I will label it *Stability-Theorist*: develop a model and then apply it with confidence, adding in disturbing factors as needed in specific cases. It was advocated by John Stuart Mill almost two centuries ago.[19]

---

[17] Earlier versions of this section appear in Northcott (2022a) and Northcott (2022b). The text here is adapted with permission.
[18] Smith (2014).
[19] Mill (1843).

A major advantage of the Stability-Theorist strategy is that it is effective even in the face of *noise*, that is, even in the face of significant effects from disturbing factors not captured by our theory. This is just as well. Outside the laboratory, no longer shielded, noise is a constant threat. For example, the moon's motion might be influenced by gravity from a second moon, by impact with a comet, or even by human interference. If so, because of these disturbing factors, the Newtonian two-body model will no longer predict accurately. Nevertheless, it still reliably identifies *one* of the factors that influence the moon's motion, namely the gravitational attraction between the moon and the planet. In this way, it still explains partially and still provides some understanding, even in the many cases where empirical accuracy is imperfect.[20] Such an achievement, advocates claim, is even *superior* to mere empirical accuracy. Why? Because empirical accuracy in any given case requires taking account of every local factor, no matter how sui generis or transient. If we wish to learn those factors that generalize, we must abstract away from the sui generis and transient, which is just what a master model does.

The Stability-Theorist strategy relies on predictability. A master model is a reliable base onto which case-specific disturbing factors can be added, only if the relations it describes operate reliably. In easy cases, warrant to apply a master model comes from empirical success here and now: the Newtonian gravity model, for example, gains warrant by successfully predicting the motion of the moon. But often there is no empirical success here and now, because of noise. Then, warrant can come only indirectly, by importing it from empirical success elsewhere, as discussed in Chapter 3. For example, even when noise means it predicts badly here and now, still we may justifiably claim that the Newtonian model correctly identifies one force at work, because of that model's empirical success elsewhere. But this inference relies on stability. The Newtonian model's warrant from success elsewhere stays good over here only because it may safely be assumed that gravity is still operating in the same way.

In sum, regardless of noise, the crucial thing for the Stability-Theorist strategy is that a relation operates predictably. *Without* noise, and with stability, a master model will be empirically accurate across many cases. *With* noise, meanwhile, while a master model is no longer empirically accurate, with stability we may still be confident that it captures some of the factors present, even if additional disturbing factors are present too.

The underlying methodological attraction is *efficiency*. Stability means a master model may be applied "off the shelf" to new cases, and models derived

---

[20] Northcott (2013b).

from it may immediately be assumed to capture real-world relations. This short cut is fantastically useful.

4.4.2 Case-Worker strategy

Things are different, though, when relations in a theory hold *un*reliably. Imagine now a different moon example, in which this time the "moon" in question is a toy moon on a string, carried around a toy planet by a child. How might we predict the motion of *this* moon? The best candidate here for a master model is presumably something psychological, perhaps that children continue doing what they enjoy. This model predicts that, if a child has carried the toy moon around the planet for two "orbits" happily, they will continue for at least another two orbits. Sometimes, that prediction will be right. Sometimes, it will not. Perhaps the child will get distracted, interrupted, or bored, or perhaps they are following instructions in an online science class (two orbits only), or perhaps they are playing a game with a friend (take it in turns to hold the moon). The model behind the prediction of continuity holds unpredictably, so relying on just that model is not effective.

What alternative strategy works better? There are many different models of children's behavior, some relatively formal, others more akin to loose hypotheses or rules of thumb that cover behavior in each of the deviant scenarios above: when a child is distracted, bored, in a school class, with a friend, and so on. Other models cover further scenarios, such as when a child is tired, or with a sibling, or affected by poverty, divorce, or a move to a new house. The key is which of these many models applies in any given case. To discover that, we need to look for contextual clues and triggers: the character of this child, the nature of this household, is the child tired late in the day, is the child hungry, is the child—or friend or parent—frustrated after a prolonged lockdown, is the weather bright and warm or is it gray and miserable, and so on. In short, exactly the things a parent considers when they try to explain or predict a child's behavior. Instead of a single master model, choose from many different models, case by case.

Label this the *Case-Worker* strategy. Case-Worker is not against wide-scope models; on the contrary, the larger the toolbox of such models, the better. Rather, Case-Worker implies two things. First, a change in balance: because no single model may be assumed always to apply, more effort must be devoted to local investigation to see which model *does* apply. Second, a change in how models are developed—discussed in more detail later in the chapter.

Sometimes, unlike in the Newtonian gravity case, we do not derive interventions, predictions, and explanations directly from a theory or model. Instead, as with the spectrum auction, the eventual intervention emerges from a tangle of different features, many of them informal, and borrows from many different models. Such heuristicist cases require extensive case-specific investigation; they show the Case-Worker strategy in action.

To clarify: with both Stability-Theorist and Case-Worker, we use wide-scope models, and contextual work is required to estimate parameter values within those models. The difference does not lie in that; it lies elsewhere. In the moon example, the moon's position and velocity vary continuously. The Newtonian gravity model tells us not just the details of that variation *but also when to expect it*. There is no "surprise" variation in gravity's influence that requires knowledge from beyond the model to predict: the inverse-square law itself does not vary unpredictably, and gravity is present reliably. Stability-Theorist therefore works well: the same Newtonian model may safely be re-applied time after time. But in many other cases, we do get just such surprise variation. Then, we *further* need to investigate each time not just a model's parameter values, but also whether the model even applies at all—whether its relations have changed their forms, or whether its relations any longer hold in any form. That mandates Case-Worker, not Stability-Theorist.

When a relation holds unpredictably, warrant requires empirical accuracy here and now. Such warrant, though, becomes harder to attain: the toy moon's motion is harder to predict than is the real moon's. But that, so to speak, is nature's fault, not ours. Still, this is not a counsel of despair: we can get a decent grip on the toy moon's motion sometimes, some predictions are better than others, and some explanations are fuller and more warranted. It is up to us to find them.

In sum, what matters methodologically is how predictably explanans relations hold. This generates the core divide. If relations do hold predictably, we may rely on a master model (if we have a good enough one), such as the Newtonian model of gravity. That is Stability-Theorist. It is effective even when empirical accuracy is disrupted by noise. But if relations do not hold predictably, a shift of emphasis is required. Now, contextual, historian-like sifting is needed each time to discover which of many candidate models applies, and any choice needs to be warranted by empirical accuracy here and now. That is Case-Worker.

A sharp dichotomy between Case-Worker and Stability-Theorist lies at the heart of the definition of fragility, and thus at the heart of the book. No doubt, research programs can feature complex mixtures of the two strategies, and individual researchers sometimes switch between them, perhaps bringing in different models used differently. But the core divide remains.

4.4.3   A familiar divide

The core divide does not come out of the blue. In addition to the case made in this book, support for it comes from how often it has arisen elsewhere. I am not alone. The divide appears whenever—often in a domain that is unshielded or complex—it is hard to achieve both generality and empirical accuracy at the same time. How, in such circumstances, should we theorize? What is new in the book is not the divide itself but the response to it.

Perhaps the most notable precursor is a famous debate from the birth of modern economics. The "methodenstreit" in the nineteenth century pitched the Austrian School of Carl Menger against the (German) Historical School of Gustav Schmoller. Although there were policy and other overtones, the heart of the debate was methodological: how best should the new science of economics be pursued? Schmoller emphasized the use of historical evidence and the need for (in our terminology) empirical accuracy. He disdained any dream of reducing human sciences to laws developed in the abstract—in his eyes, this dream is a red herring that harms science by distracting us from close empirical work. Menger, by contrast, argued that human actions and interactions are too complex to allow for useful empirical regularities. An emphasis on empirical accuracy therefore dooms us to idiosyncratic historical case studies with no generalizability. Better to develop theory a priori. In the case of economics, that means developing theory around an assumption that agents pursue their economic self-interest.

The analogy is obvious. Various details notwithstanding, Schmoller advocated for a version of Case-Worker, Menger for a version of Stability-Theorist. These are the two distinct options. Ever since, methodological debates within economics have repeatedly landed across this same fault line. Most recently, the "empirical turn" has been an empiricist reaction against the perceived dominance of a priori rational-choice theory (see Chapter 8).

It is not just economics. Versions of rational-choice theorizing have spread widely in sociology, political science, and other social sciences. In all cases, there has been an empiricist reaction.

Further, it is not just social science. A similar divide appears in natural science, too. For example, it has been argued that two distinct traditions have informed the study of mass extinctions: a phenomena-led approach from paleontology versus a theory-led analysis of major transitions in natural history.[21] Again, this is, roughly, Case-Worker versus Stability-Theorist.

Finally, it is not just science. Our divide recalls the divide in early modern epistemology between empiricism and rationalism (some disanalogies notwithstanding). Should we follow every empirical toss and turn, or should we stick to certain commitments that are, so to speak, pre-empirical? In the book, these two classic positions are represented by alternative research strategies, namely Case-Worker and Stability-Theorist, each of which is suitable in some cases but not in others.

4.5   The importance of case studies

I now develop the methodological implications of the core divide. To begin: how might we implement the Case-Worker strategy? One important method is to learn by applying theory to real-world cases—in other words, to learn from case studies.

With case studies, is our goal theory development, with case studies merely aids to that? Or is it the other way around, with our goal being the successful analyses of individual cases, and theories merely aids to *that*? I think both goals are important. But either way, theories are important, so we need to understand how case studies contribute to theory development—because for that they are crucial.

A naive, but nonetheless common, view is that case studies are tests of a theory's truth: if a theory explains or predicts a case well, it is confirmed, if it does not, it is falsified. But this view implicitly assumes stability. If a relation is fragile, by contrast, then typically it will not hold everywhere in a domain, so any theory built around that relation will not hold everywhere, either. Case studies are not crucial experiments that dictate a verdict for all contexts. (Indeed, neither always are experiments. As the "new experimentalist" philosophy has emphasized, experiments play many other roles too.[22]) We need a more sophisticated understanding.

---

[21] Currie (2019b).
[22] Hacking (1983); Weber (2018); Franklin and Perovic (2021).

Several complaints deny that case studies can help with theory development.[23] A first such complaint is that case studies are plagued by research designs that are too informal and have too many free variables, so we cannot confirm causal explanations even for single cases. In reply, no doubt there are examples of bad practice, but as a general thesis, this pessimism is excessive, not to mention philistine. It does not stand up to the details of actual case studies; it ignores a large literature on causal process tracing[24]; and as I discuss in Chapter 6, upon closer inspection, supposedly more rigorous methods of causal inference themselves require supplementary assumptions that rely on less formal methods.

Other complaints target, so to speak, bad external rather than internal validity. One complaint insists that any type-level knowledge must be inferred from a large sample, and a case study represents a uselessly small sample of 1. In reply, as I discuss in Chapter 6, large-sample statistical studies are suitable only when relations are stable. When they are fragile then, by contrast, investigation must be local.

A related complaint is that the result or lesson of a case study is nonreplicable: a discovery from one case typically does not extrapolate to others. So, it is claimed, case studies cannot help science progress. In reply, indeed no algorithm exists for extrapolating the results of case studies, and case studies cannot establish statistical generalizations. But to think these flaws are fatal is to be in the grip of a simplistic picture of theoretical progress.

Turn to a positive account. First, a case study potentially brings two distinct benefits. The first benefit is simply to illuminate the case at hand, usually by elucidating the case's detailed causal structure. To this end, case studies often employ multiple methods: quantitative, but also interviews, participant observations, and consultation of archives or official datasets. Ashworth, for example, interviewed veterans and consulted regimental archives, mixing both qualitative and quantitative analysis, to establish a full picture of the World War One truces.

But just as critics charge, only rarely may an explanation be assumed to extrapolate to other cases, because individual cases differ too much. Case studies must offer more. And they do. Turn now to their second benefit: to develop theory.

---

[23] For references, see Ylikoski (2019), who has also influenced much else in this section. For more general discussion of case studies, see Morgan (2012); Ankeny (2012); and Ylikoski and Zahle (2019).
[24] Collier (2011); Runhardt (2015).

Some ways in which case studies develop theory are familiar: they might show a theory applying well, or a theory not applying well, or a situation not well covered by any current theory. But while such findings are useful, none gets to the heart of case studies' true value here.

Direct one-to-one mappings between theory and explanation, in the manner of Newtonian textbooks, do occur. Often, though, causal explanations draw on many different theories, as when Ashworth's explanations of World War One truces drew on many different psychological mechanisms. An eventual explanation or intervention might not be derivable from existing theories at all, as with the spectrum auction. Instead, theories provide building blocks. Theoretical progress is then a matter of expanding or refining this toolbox of building blocks, and case studies' most important contribution is to help with that. They do so in several ways.

4.5.1 Suggest new theory

Case studies can, most simply, suggest new theory by revealing a causal pattern or structure that might apply more widely. This technique is ancient. The Greek historian Thucydides used the case study of the civil war in Corcyra to develop theoretical claims about civil wars in general.[25]

4.5.2 Refine existing theory

Case studies are essential to the *refinement* of theories, a process at the heart of science. Petri Ylikoski gives the example of the self-fulfilling prophecy—the theory in sociology that the mere prediction of an effect itself causes or amplifies that effect. One case study looked at the impact on US law schools of the introduction of public rankings.[26] It revealed several distinct ways in which a self-fulfilling prophecy occurs. First, the initial ranking influences later rankings via a reputation effect beyond any data the initial ranking was based on, because later rankers rank in part according just to reputation, which is influenced by earlier rankings. Second, an initially chancy or insignificant ranking difference nonetheless impacts application numbers, which changes the scores on various metrics, which in turn impacts later rankings. Third, a

---

[25] Thucydides (1974/c 400 BCE).
[26] Espeland and Sauber (2016), discussed in Ylikoski (2019).

school's rank influences how administrators allocate resources, which impacts the school's subsequent ranking. Fourth, schools are incentivized to game the rankings by prioritizing those aspects of their provision that the rankings emphasize, which itself reinforces the validity of the rankings' measure. These different mechanisms by which the rankings become a self-fulfilling prophecy all potentially apply to other cases too. Thanks to the case study, the theory of the self-fulfilling prophecy is thus sharpened for use elsewhere.

These new mechanisms are, in effect, themselves new theory. The refinement of existing theory and the development of new theory sometimes overlap and can be hard to tell apart.

How details of a theory are tweaked to apply to the case at hand, and how a theory interacts with an additional theory that is brought in—to master that, there is no substitute for *using* theory in actual cases.

Because case studies tease out underlying mechanisms, they are emphasized by critical realists, for whom identifying underlying mechanisms is one of social science's major goals. The intensive nature of case studies is claimed to enhance researchers' sensitivity to such mechanisms.[27]

4.5.3 Establish diagnostic indicators

Another crucial task for which case studies are essential is to establish diagnostic indicators: when should we expect that a theory applies? Nancy Cartwright, in recent work, calls such indicators *markers* (in the positive case) and *derailers* (in the negative case).[28] Sometimes, markers are just correlates or effects of a mechanism operating, other times they are necessary support factors. Similarly, derailers may be correlates or effects of a mechanism not operating, or they may be causal blockers.

If a relation is fragile, we do not know in advance whether it holds. Supplementary investigation is required to find out, and a large part of such investigation is tracking whether diagnostic indicators are present.[29] That, of course, requires knowing what the diagnostic indicators are. And in practice, what they are is often learned only from the experience of applying the theory, that is, from case studies.

---

[27] Ackroyd (2009).
[28] Cartwright (2020a, 2021); Cartwright et al. (2022).
[29] Mitchell (2000).

Generally, case studies are the route to "learning by noticing."[30] By applying a theory or model to an actual case, a scientist learns about factors they had previously ignored or not thought of but that turned out to be crucial.

4.5.4 Nontheoretical progress

As well as developing theory, a case study can also lead to advances in methodology, to new concepts, and to new research questions. But here, I retain a focus on theories.

In sum, case studies suggest new theory, reveal how theories combine, and reveal the limitations of existing theory. They flesh out how theories may be applied: they reveal different mechanisms through which they apply, they establish what a mechanism's support factors and blockers are, and they establish indicators of whether a mechanism has or has not been operating. Such work is the key not just to empirical progress but also to *theoretical* progress. When relations are fragile, theories cannot easily be developed without case studies. No other method is as good.

4.6 Middle-range theories

Case studies are essential not just to science; they are essential also to our philosophical understanding *of* science. They give evidence that fragility is widespread. They also reveal the role of *middle-range* (i.e., middle-level) theories. These are theories that, first, are of wider scope than simple empirical hypotheses that arise in everyday research, and yet second, are also of narrower scope than grand, high-level theories.

Middle-range theories were first suggested as a category by Robert Merton, and recent philosophy of science has emphasized their role in generating explanations and interventions.[31] In Ashworth's explanations of the World War One truces, for example, several psychological mechanisms are crucial, but these mechanisms are not one-off causal relations, because they apply in many cases. They are not grand, high-level theory, either. Rather, they are middle range.

---

[30] Duflo (2017).
[31] Merton (1967, 1968); Hedström and Ylikoski (2010); Elster (2015); Cartwright (2020a).

The benefits of case studies for theory development also tend to be at the middle level. In the law school rankings case, the new mechanisms via which the self-fulfilling prophecy happens are all middle range.

4.7 The core divide again: two ways to develop theory

Just as the best strategy for *applying* theory depends on how predictably a relation holds, so too does the best strategy for *developing* theory. Reflecting this common dependence, the same methodological dichotomy reappears.

This reappearance should not be surprising. Sometimes, the terms "develop" and "apply" refer to much the same thing anyway. A Newtonian force model of a new bridge design, for example, might equally be thought of as either an application of Newtonian theory or a development of it. Despite this overlap, though, it is useful to distinguish the two concepts. In Chapter 8, for example, I discuss work in economics that is commendably Case-Worker in its theory application but unfortunately Stability-Theorist in its theory development, and we want to be able to diagnose that.

4.7.1 Case-Worker again

We have learned several lessons for when relations are fragile. First, to develop healthily, a theory needs constant empirical refinement. Second, useful theories are usually middle range. These two lessons are connected: middle-range theories are used in applications more easily than are high-level theories, and thus they are empirically refined more easily. A third lesson is the value of markers of when a theory applies.

What strategy for theory development follows from these lessons? The answer is described well by the late sociologist Howard Becker:

My work doesn't produce timeless generalizations about relations between variables. It results instead in the identification of new elements of a situation, new things that can vary in ways that will affect the outcome I'm interested in, or new steps in a process I thought I'd understood until a result different from what I expected occurred. I can use these new elements of organization and process to direct my next inquiry. For me, that's the way social science works.[32]

This is a version of Case-Worker. To know when a relation or combination of relations works reliably, we need experience of trying. Only in this way can we keep a theory empirically fruitful as we develop it.

4.7.2 Stability-Theorist again

Contrast this to a very different paradigm for theory development: abstract mathematical derivation, after the model of Euclid's *Elements*. This paradigm is well suited to mathematics, whose relations hold with certainty—mathematical relations are stable par excellence. The same paradigm can be applied to the physical realm too, and indeed it has been—with great profit—in some of science's most famous episodes.

Consider Newtonian mechanics. The Newtonian laws that (approximately) govern force interactions are universal, so models of new situations, be they bridges, pendulums, or rotating spheres, may be developed using those same laws, in full confidence that they still apply. We do not need continuous testing. This is a wonderful shortcut. It is justified by the models' eventual empirical success—empirical confirmation is still there, but for some periods we may leave it on autopilot, so to speak.

Refinement is made easier, too. If a model predicts wrongly, an adjustment—always informed by the assumption that Newtonian relations still hold—is often available immediately: perhaps a mismeasured force here or an omitted force there. Again, the endorsement of this strategy is ultimately the empirical success it leads to.

There are many other inputs to engineering practice beyond basic Newtonian models, of course. The point is that engineering's empirical success endorses how Newtonian theory has been developed, relative to developing it without Stability-Theorist short cuts.

Once the fundamentals of a theory are empirically vindicated, Stability-Theorist may develop that theory rapidly, confident that if the building blocks and the rules for composing them operate reliably then so will what we construct from them. Mathematical derivation may run free.

---

[32] Becker (2014, 3), quoted in Ylikoski (2019).

4.7.3   When Stability-Theorist goes wrong

Here is the danger: if a relation holds only unpredictably, Euclid is no longer a helpful exemplar, and the Stability-Theorist strategy breaks down. A new relation derived from relations that hold unpredictably often itself holds unpredictably, and if new relations after each round of derivation are not endorsed by empirical refinement each time, the potential for decay multiplies, risking drift into a fantasy-land house of cards. Empirical validation may no longer be left on autopilot.

Consider the career of the Prisoner's Dilemma (see Chapter 3). This game has been developed almost entirely in the Stability-Theorist mode, insulated from empirical refinement, new versions instead derived mathematically from foundational principles of rational-choice equilibria. A "casual empiricism" does motivate many developments, such as versions with more than two agents or with agents with incomplete information. But this is not the same as true empirical refinement. There are almost no case studies, and thus none of the crucial benefits those bring: modifying the game's assumptions to make it more applicable, exploring sub-mechanisms by which the game's strategic logic acts out in actual cases, or learning what markers track whether the game applies well. None of this would be a problem if the relevant relations held reliably. In that case, we could derive different Prisoner's Dilemma models from fundamental principles just as we derive different Newtonian models. But real humans do not reliably behave like the rational actors in a Prisoner's Dilemma model, as overwhelming evidence shows. As a result, the many intricate turns of the literature have not proved fruitful empirically.

The Prisoner's Dilemma is no outlier. A chronic problem with "toy" models—that is, with models that are highly simplified and abstracted—is that they are not developed in an empirical, Case-Worker way. Similar criticisms apply to much rational choice modeling generally, in economics, in other social sciences, and in mathematical ecology.[33] They also tell against the frequent philosophical support for field sciences to develop models via an internal logic rather than via case-to-case empirical refinement.[34]

---

[33] Northcott (2018); Northcott and Alexandrova (2015); Farmer (2013); Sagoff (2016).
[34] Notable examples of such support include Elster (1989); Little (1991); Lawson (1997); and Brante (2001).

4.8 Example of theory development: political violence

Consider now an exemplar of Case-Worker theory development.[35] Donatella della Porta's 1995 book, *Social Movements, Political Violence, and the State*, is a highly influential study of political violence in 1960s and 1970s Italy and Germany.[36] The book is known for emphasizing sociological over ideological or psycho-pathological explanations—terrorists are not just evil fanatics.

Della Porta's primary goal is to causally explain, and to this end, she adopts, in effect, the Case-Worker strategy. Despite this, her book is famous for being innovative theoretically. How so? Like game-changing work generally, she establishes new mechanisms and categories that successors are obliged to consider—new tools for the toolbox.

A bedrock of della Porta's method is the willingness to use different theories whenever these pay their way with new causal explanations. Della Porta employs mixed methods in a similar spirit. Individual actors' life histories, in the form of qualitative analysis of interviews, form part of her evidence base, but so too do quantitative data, such as statistics about acts of violence. In addition, she uses archival research, such as consulting official records. These different kinds of evidence support different inferences in the service of larger explanatory ambitions.

Consider one of those larger ambitions: what explains violent groups' behavior? Some previous work focuses on broad political determinants, such as the scope within a polity for expression of political frustrations. Other previous work focuses on rational-choice modeling of what tactics best achieve a group's ideological goals. Still other previous work, and popular coverage, focus on individual-level psycho-pathology—actors might be deemed evil or psychopathic. Della Porta deviates from all of these. She instead examines organizational dynamics at the group rather than society level. She finds that these dynamics are "irrational" in the sense that they are not driven by groups' ostensible ideological goals, but she also finds that individual actors are usually "rational" in the sense of not being pathologically violent or immoral.[37]

Her analysis begins with arrests by police. These arrests disproportionately weaken those groups that are organized loosely, which creates a selection effect in favor of groups that are more centralized and compartmentalized. This leads to reduced recruitment of new members, and so to the dominance of

---

[35] This section is adapted with permission from a part of Northcott (2023).
[36] Della Porta (1995). Over 1,500 citations, according to Google Scholar.
[37] della Porta (1995, 116–33).

internal concerns. Actions are chosen to achieve internal goals such as discipline or self-defense (robberies, shoot-outs during arrests, punishment of "traitors") rather than, as earlier, external goals such as propaganda or campaigns (targeting of unpopular factories or businesses). The emphasis on self-defense rather than recruitment leads tactics to become increasingly lethal and bloody. Ideology evolves accordingly and becomes incomprehensible to outsiders, with less emphasis on propaganda for external consumption and more emphasis on internal cohesion. The more underground and sealed off a group becomes, finally, the less effectively it influences wider society.

Throughout della Porta's book, theory is developed and sharpened by detailed engagement with her Italy and Germany case studies. No abstract mathematical derivations, and hence, no corpus of theory left bereft of explanatory traction. Instead, new mechanisms and categories are developed only when they are empirically fruitful. This is Case-Worker in action.

One example of a new category is the *policing of protests*.[38] Police tactics serve as a downstream proxy for deeper state factors and institutional features, such as police organization, the nature of the judiciary, law codes, and constitutional rights. This simplifies tracking of the state's influence on the (already theorized) political opportunity structure because the connection between policing and social movements is conveniently direct. It also enables policing itself to be analyzed in a subtler way than before. New explanations result. Police tactics, for example, became more hardline not because of internal dynamics within the police but rather because of external political decisions that the police tried to resist, contrary to much previous theory.[39] Other new explanations include: how hardline state and police attitudes rose and fell with the attitude and strength of the moderate "old left"; how political polarization strengthened the hands of hardliners on both sides; how, in the long run, hardliners declined in influence; and how the tactics of the protestors influenced the tactics of the police.[40]

This rich explanatory detail is made visible by della Porta's new theories. These are not formal models; rather, they are qualitative and verbal. Her use of these theories is sometimes heuristicist rather than causalist, as rather than specify causal hypotheses directly they instead bring into view new categories or ways of seeing things. So is her use of others' theories (italics added): "recent studies on social movements *provide the main categories* for the explanatory

---

[38] della Porta (1995, 56).
[39] della Porta (1995, 77–8).
[40] della Porta (1995, 76–8).

model of political violence in Italy and Germany that I am going to develop here."[41]

Della Porta is explicitly opposed to universal theory or wide-scope explanation.[42] The causal relations that govern political violence hold unpredictably. Whether and how they hold is sensitive to many variables: leftist versus rightist protest movements; democratic versus authoritarian political environments; class versus ethnic bases; different organizational models, forms of action, and ideologies and goals; and no doubt other things besides. Della Porta takes her own work to apply reliably only to leftist, class-based groups in a democratic environment. At the end of her book, she cautiously examines how well it extrapolates to the civil rights movement in the 1960s and 1970s in the United States.[43] Because of fragility, this extrapolation cannot be assumed—and indeed, della Porta concludes, it fails. The point is that a supplementary investigation is needed to find that out.

Some relations recur across many contexts, and knowledge of them is essential—this is theory's familiar role. But theory must be developed effectively. When relations are fragile, that means avoiding Stability-Theorist: no grand but useless superstructures. Instead, Case-Worker is required. For a theory to be sharp and useful, it needs continuous empirical refinement. Della Porta's work exemplifies that, which is why it succeeds.

4.9 Fear of missing out

How can we tell whether a relation is fragile? There is no failsafe algorithm. In practice, though, we do not need one, because whether a relation holds reliably usually becomes clear soon enough. In the case studies in this book, for example, it is not controversial that none of the rules of thumb about invasive species hold reliably (Chapter 5), that iterated interaction across the trenches did not lead to truces reliably (Chapter 3), or that many pandemic policy interventions did not have reliable impacts (Chapter 10).

Background knowledge and experience prime us. As we will see, fragility seems more common in biology than in physics, and more common still in social science—consistent with folk wisdom, such as Aristotle's admonition against expecting too much precision in analyses of human affairs. Relatedly,

---

[41] della Porta (1995, 9).
[42] della Porta (1995, 210).
[43] della Porta (1995, 210–5).

fragility seems more common in systems that are complex and in systems that are unshielded.

There exists a formal analysis of when, and to what extent, absence of evidence becomes evidence of absence.[44] In our case, when would the absence of evidence for stability become evidence of stability's absence? The formal analysis is of limited value for our purposes, though, because filling out the probabilities in the formulas cannot be done without investigation on the ground, after which whether a relation is fragile is usually obvious anyway.

Once investigations begin, stability has a tell-tale symptom: reliable interventions. We can, for example, reliably accelerate a ball by striking it, because the relevant relation is stable. Even if the environment is noisy and the ball's overall acceleration is hard to predict, still we may be confident about a strike's incremental effect. Thus, we have a marker of stability: does a relation license reliable interventions? Can we *do* things with it?

But there are deeper issues at play. What often lies behind the inquiry about how to tell whether a relation is stable? Fear—of missing out. Discovering a stable relation is tremendously useful; do we risk missing out on that if we settle prematurely for Case-Worker? Better, on this view, to err by presuming stability too much than too little.

This fear, though, is misplaced, even dangerous, and it is important to see why. First, in practice, the erring is almost all the other way around. There are many examples of Stability-Theorist being pursued when Case-Worker should be, as discussed throughout the book, but, at least in cases I have seen, few or no examples of the opposite mistake. A scientist is usually delighted to discover a stable relation, and they will not easily veer away from trying.

Second, mistakenly pursuing Stability-Theorist rather than Case-Worker is costly. In all, 29,000 Prisoner's Dilemma articles, for example, represent a huge opportunity cost of other, more productive research foregone—many potential interventions and explanations passed up. That is the price of "one more heave."

Third, when an environment is noisy and full of fragile relations, Case-Worker is in any case usually the best way to ferret out whatever nuggets of stability do exist. The invasive species case study in Chapter 5 illustrates. To preview that: for the most part, no stable relations have been found that govern whether a species invasion succeeds, but an exception is a set of species of pines for which reliable predictors *have* been discovered. How were they discovered? By Case-Worker. The intricate modeling and measurement required were

---

[44] Sober (2009).

developed and applied empirically at all stages. *After* being discovered, these stable relations could then be exploited by Stability-Theorist—but only then.

The lesson generalizes. Stability-Theorist is effective only when used with a relation that is stable, and in field situations often we do not know in advance whether a relation is stable. If so, then we cannot reliably predict when the relation holds, and so, by definition, the relation is fragile, which means initial investigation is best done with Case-Worker. So, the fear is turned on its head—when uncertain, to ensure we do not miss out on any stability, *Case-Worker* is required first, not Stability-Theorist.

In sum, usually, we can tell relatively quickly whether a relation holds reliably. A head-in-the-sand, one-more-heave insistence on Stability-Theorist, meanwhile, carries grave costs in the many cases of fragility, while it also, ironically, makes us *more* likely to miss out on what stability there is.

4.10 Cartwright and others

How do the ideas in this book relate to the work of others? One obvious, big influence is Nancy Cartwright, and more broadly the "Stanford school" of philosophy of science. Recent work by Cartwright and others dovetails with, and inspires, many of my own arguments.[45]

Famously, Cartwright was an early skeptic of scientific realism about theories.[46] Later, she criticized what she called the "vending machine" view of theory. According to the vending machine view, roughly, theories may be applied purely using their own resources, with no additional need to develop models of local circumstances or to bring in other contextual knowledge. In most cases, Cartwright argues, this view is false.[47] In more recent work, Cartwright and co-authors argue that actionable scientific knowledge requires a "tangle" of extra-theoretical ingredients: models; measurement definitions, procedures, and instruments; institutions; concept development and validation; data collection, analysis, and curation; statistical techniques; methods of approximation; narratives; bridging principles; classification schemes; practices; devices and materials; equipment; model organisms; know-how; and much else.[48]

---

[45] I have in mind especially Cartwright and Hardie (2012); Cartwright (2019, 2020a); and Cartwright et al. (2022).
[46] Cartwright (1983).
[47] Cartwright (1999, 58–9).
[48] Cartwright (2020a, 2020b); Cartwright et al. (2022). See also Boumans (2015).

When criticizing the vending machine view, Cartwright had in mind theories comprised of stable relations, such as classical and quantum mechanics. In her view, to apply even these theories requires a tangle. I agree. But as Cartwright herself advocates, that need not tell against Mill's strategy of applying the same master theory widely, adding in a suitable tangle with each new application. It remains efficient to devote resources to developing such a master theory, so we should endorse Stability-Theorist when relations are stable.

In this book, I extend Cartwright's themes from, so to speak, the stable world to the fragile one. Cartwright's own preferred ontology is causal capacities, which are conceived to hold stably. Fragility, meanwhile, is a property of relations, including the causal relations that constitute a capacity. So, capacities may be fragile, so to speak—and often are.[49] Case-Worker follows. It dovetails with the tangled view of science, which already suggests that resources and prestige should shift from theorist to case worker.

One advantage of Cartwright's ontology of capacities is that capacities may be combined in endlessly different ways. Sometimes, this arrangement of capacities, rather than the capacities themselves, is what holds unpredictably. The distinction matters. If the operation of capacities, or of relations generally, is predictable and only their arrangement is not, then the Stability-Theorist strategy remains desirable: we are well served by developing stable building blocks and stable ways to combine them, as if developing a Meccano toy set.[50] The extra focus in this book is unpredictability that stems from the capacities themselves, not just from arrangements of them.

In a nutshell, in Table 2.1 (from page 21) while Cartwright's work has focused on the top-right slot, I focus on the top-left one.

This book is organized around the unifying thread of fragility. Although no previous work makes fragility its centerpiece in the same way, much has, in effect, appealed to similar ideas. Cartwright herself uses the word "fragile" to describe causal relations that hold unreliably because they are highly sensitive to changes in underlying causal structure—the objective face of fragility.[51] Such relations are common in policy contexts, she argues. To exploit them requires

---

[49] I use the notion of a fragile capacity here to clarify the relation between my work and Cartwright's. But note that in Cartwright's system, capacities are stable by definition, so a "fragile capacity" is an oxymoron. A causal disposition that holds only unreliably, she would instead describe as a "tendency principle" or some other term. (I thank Nancy Cartwright for clarifying this point.)

[50] I take the analogy with Meccano from Cartwright et al. (2022, 151–5).

[51] Cartwright (2012, 977).

knowledge of that underlying structure, which in turn requires the supplementary investigation characteristic of Case-Worker.

Sandra Mitchell, like many, has long emphasized that generalizations in biology hold unreliably. How to find out when they do hold? "To know when to rely on a generalization that does not apply to all space and time we need to know when it will apply, and this can be decided only from knowing under what specific conditions it has applied before."[52] This anticipates the discussion earlier of markers and case studies. What is the root of the difference between laws in biology and laws in physics? According to Mitchell, it is stability: "the difference between the laws of physics, the laws of biology, and the so-called accidental generalizations is better rendered as degrees of stability of conditions upon which the relations described depend."[53] This is close to the distinction between stable and fragile relations, albeit, as often in previous work, it alludes only to fragility's objective face.

Julian Reiss has criticized the notion that economic relations are stable across different environments. He writes of "feeble facts."[54] This, along with several other of Reiss's writings, is an inspiration for my own thoughts about fragility. What Reiss has in mind by feeble facts, though, at least in many cases, is what in Chapter 2 I label contextual rather than fragile: a relation highly sensitive to background conditions, but where we have a good idea of what those sensitivities are, and so we can predict reasonably well when the relation will hold.

Jon Elster has long promoted the use of social mechanisms. He writes (italics in the original), "mechanisms are *frequently occurring* and easily recognizable causal *patterns* that are triggered under *generally unknown conditions*."[55] Triggering conditions being unknown is at the heart of fragility. Elster, though, does not develop the methodological implications of this, as his focus is more on truth and confirmation.

Adrian Currie and Kirsten Walsh have outlined a framework that endorses something like the Case-Worker strategy, depending on the nature of the target system.[56] For them, the key properties of a system are "noise," by which they mean shocks external to the system, and "interference," by which they mean laws that interfere with each other. The analogy is not perfect. But roughly,

---

[52] Mitchell (2000, 259).
[53] Mitchell (2000, 257).
[54] Reiss (2019).
[55] Elster (2015, 27).
[56] Currie and Walsh (2018).

Currie and Walsh advocate a theory-based approach akin to Stability-Theorist when either noise or interference in a system is low, and a contextual approach akin to Case-Worker when both are high.

Currie and Walsh do note the agent relativity of their noise and interference, although they do not emphasize it. Generally, while many authors recognize, in effect, the objective face of fragility, the subjective face is usually not emphasized. Yet it needs to be. Only in that way can we draw out the core methodological divide between Case-Worker and Stability-Theorist.

The program of this book is akin in spirit to many wider trends in philosophy of science. The holist tradition has long emphasized nontheory as well as theory aspects of science.[57] So too, has the theme of "science without laws."[58] This has gone hand in hand with the explosion of interest in scientific models. It also goes hand in hand with the burgeoning new mechanist literature (see Chapter 6). Not all of this work embraces fragility, but its emphasis on narrower- over wider-scope entities is congenial to it.

Pragmatism, too, distrusts theory-centered analyses of science, preferring to emphasize process, activity, and reliability.[59] Like pluralism, it is suspicious of theory monism.[60] This suspicion dovetails well with a fragility perspective, and it is directly implied by a toolbox view of theory.

Other related trends apply, roughly, a localist sensibility to venerable debates within philosophy of science. John Norton's "material theory" of induction rejects any a-contextual, formal account, and instead emphasizes that inductions are warranted by background facts, exactly which background facts varying case by case.[61] If so, all induction is local. So too, confirmation—according to Cartwright and co-authors. They argue that applying any theory of confirmation implicitly presupposes a tangle of procedures and assumptions, even when the relation being confirmed is stable.[62] Such tangles typically involve many fragile relations, and thus, so does confirmation.

---

[57] Kuhn (1962); Laudan (1977); Chang (2022); Massimi (2022).
[58] Cartwright (1983); van Fraassen (1989); Teller (2001); French (2020).
[59] Chang (2012, 2022); Potochnik (2017).
[60] Ludwig and Ruphy (2021).
[61] Norton (2021).
[62] Cartwright et al. (2022, 132–42).

4.11 Where have we reached?

We have covered a lot of ground. To recap: even when warrant for explanations is narrow scope because relations hold unpredictably, still wide-scope theories are essential. They play a toolbox role. Sometimes, they contribute to scientific success only indirectly.

When relations hold unreliably, two core errors are common: (1) claims of explanation and understanding that lack warrant and (2) theory developed abstractly, that is, without continuous empirical refinement.

When relations hold reliably, these two errors are less salient. Warrant imported from elsewhere may justify claims of explanation and understanding even without empirical confirmation here and now, and even theory developed abstractly often explains and predicts successfully. Label by *Stability-Theorist* the strategy that develops and applies theory on the assumption that a theory's relations hold reliably, so that empirical refinement and confirmation may for periods be left on autopilot. The master models and theories that Stability-Theorist delivers greatly heighten scientific efficiency.

Everything changes when relations hold only unreliably, though. Stability-Theorist becomes a mistake. Why? Because it leads to the two core errors above. The *Case-Worker* strategy is required instead. According to it, we may not presuppose that a theory's relations hold, so supplementary investigation is required each time to warrant explanations (or predictions, interventions, or understanding). Theory developed abstractly risks quickly becoming an inefficient house of cards. Case-Worker mandates that theory instead be developed via constant empirical application and refinement.

The distinction between Stability-Theorist and Case-Worker is dichotomous: either a relation holds predictably enough to merit explanatory claims without empirical confirmation each time, or it does not. And either a relation holds predictably enough that developing theory abstractly is explanatorily fruitful, or it does not. Borderline cases exist, but the distinction remains crucial. It has a long history.

Case studies illustrate. The relations in the Prisoner's Dilemma are fragile, yet in our example, the Prisoner's Dilemma was applied in a Stability-Theorist way, so its claims to have explained World War One truces are unjustified. Case-Worker, in the form of the historian Ashworth's explanations, works better. It works better for theory development too, as illustrated by della Porta's work on political violence, in contrast to other theorizing in that domain, and in contrast to how the Prisoner's Dilemma has been developed.

What matters is unreliability regarding the relation explaining the outcome, not the outcome itself—that is, regarding the explanans, not the explanandum. As discussed in Chapter 2, if an outcome is predictable, the Stability-Theorist strategy is always favored. If an outcome is *un*predictable, though, things are still open, methodologically speaking: the unpredictability might be because of noise, it might be because our relation operates unreliably, or it might be because of both. Which of these it is makes all the difference. If the cause is just noise, then Stability-Theorist is favored, but if the cause is also (or just) an unreliable relation, then Case-Worker is favored. In Table 2.1 (on page 21), this is the difference between the top-right and top-left slots.

The top-left slot in Table 2.1, finally, is the major lacuna in philosophy of science that this book addresses. It is a *lacuna* because it has been comparatively neglected. It is *major* because the top-left slot is so widespread, which tells against much scientific practice and in favor of much other scientific practice, and correspondingly tells against much philosophical opinion and in favor of much other philosophical opinion. The rest of the book seeks to make good on these claims.

5

Ubiquity of fragility

5.1 Example of fragility in natural science: Invasive species

Does fragility matter? It does if it is common, or if it holds in crucial cases. In this chapter, I argue that we should expect fragility to be common—and not only common but also, in a sense to be outlined, ubiquitous. I then survey a range of examples to make plausible that, in fact, fragility is indeed common, including in crucial cases. I begin with a case study from natural science.

Beyond the car radiators from Chapter 1, consider a more extended example from natural science: when and why are habitats invaded by outside species?[1] Key relations here turn out to be fragile. The claim is not that *therefore* fragility is common throughout natural science; I use the case merely to illustrate that it is not restricted to social science. I report briefly on more examples of fragility in natural science in the final section of this chapter.

Species invasions are hard to predict. Even though many mechanisms behind them are known, usually we cannot predict reliably which of these mechanisms will hold. They are fragile. As a result, we also cannot predict reliably an invasion's outcome or scope. Every invasion is highly sensitive to idiosyncratic local factors: "whether or not a particular invasion will succeed depends not only on the common aspects and mechanisms shared by a number of systems, but on the peculiar ways in which all the small details of a system interact."[2]

One of Alkistis Elliott-Graves's examples is plant–soil interaction.[3] There exist both positive (certain fungi and nitrogen fixers) and negative (pathogenic microbes) potential feedback to plants from the soil. Evolutionary interaction tends to favor the negative feedback, with the result that plants tend to prosper more in a new area. Does this relation hold reliably enough to predict plant invasions by analyzing invaders and soil communities? Alas, no. How quickly plants accumulate pathogens is critical to an invasion's success, and

---

[1] This section is adapted with permission from a part of Northcott (2022b). It draws primarily from the work of Alkistis Elliott-Graves (2016, 2018, 2019).
[2] Elliott-Graves (2016, 377).
[3] Elliott-Graves (2016).

Science for a Fragile World. Robert Northcott, Oxford University Press. © Robert Northcott 2025.
DOI: 10.1093/9780191944352.003.0005

this in turn varies with several further, local factors, such as the relative abundance of invaders and native plants, and the predation climate. These are hard to predict. That is, the relation between plant–soil set-up and invasion success holds unpredictably: even when a combination of invader and soil microbes seems perfect for an invasion to succeed, often the invasion fails, nevertheless. Plant–soil feedback interactions have been modeled extensively, and the relative abundance *within* a community of all-native plants has been predicted successfully. But when it comes to invasions of new communities, predictions are no longer reliable.

Many other rules of thumb explain, or partially explain, invasions sometimes. But none holds reliably. Examples include that islands, especially small ones, are more susceptible to invasions than are mainlands; temperate climates are more susceptible than the tropics; and within a taxon, smaller animals are more invasive than larger animals. Within plant taxa, the following traits help invasions succeed: small seed size; phenotypic plasticity; allelopathy, that is, producing biochemicals that impact the success of other organisms; adaptation to fire; and, at different times in different places, small and large size, flowering early and late, and dormancy and nondormancy. And the following properties of communities help invasions succeed: the community is disturbed (i.e., there is an ongoing environmental change); lack of biological inertia (i.e., the ecological balance can change relatively easily); some plant–soil feedback, as discussed above; and, at different times in different places, both high and low diversity. Similar lists can be compiled for marine ecosystems, insects, vertebrates, and so on. But because no single rule of thumb is reliable, for any given invasion, extensive case-specific investigation is required to find out which, if any, rules of thumb explain it.

Elliott-Graves criticizes attempts to develop a grand, one-size-fits-all master model.[4] Like with the Prisoner's Dilemma, these are developed in insulation from empirical refinement, and they abstract away from idiosyncratic details of individual cases—even though those details are crucial to explanatory success.[5] As a result, resources spent on general models have not paid their way, any more than they have with the Prisoner's Dilemma: none of the general models has improved predictive accuracy or has achieved empirical warrant for its explanations.

Richard Levins famously argued that ecological models face a trade-off: they cannot simultaneously maximize realism, generality, and precision.[6] Yet this

---

[4] Edwards-Graves (2016, 380) lists a dozen examples of such attempts.
[5] Sagoff (2016) criticizes mathematical ecology generally, on similar grounds.
[6] Levins (1966).

misses that some complex systems can be analyzed with success—when parts and their interactions operate reliably, as with the many parts in an aircraft, for example. The real problem in invasion biology is not complexity; it is fragility. Ecosystems are not aircraft, and the same entity can have different effects in different invasions in ways that are hard to predict. Every invasion is idiosyncratic. Slight differences can make all the difference to the outcome, so that even when the sources of two invasions seem very similar, extrapolating from the result of one to the other is unreliable.

How, then, might progress be made? A rare oasis of predictive success shows the way forward: the characteristics that distinguish invasive from noninvasive species of pine.[7] There exist unusually good quantitative data for pines, and for conifers generally, including for failed invasions (which for other species are often not reported) as well as successful ones. Based on this data, 24 pine species were classified as invasive or not. What traits would correlate with this classification? Marcel Rejmánek and Da vid Richardson ran a statistical discriminant analysis on 10 traits suggested by background theory. An "unusual robustness of classification" identified three of these traits as individually necessary and jointly sufficient for indicating which pine species are invasion threats and which are not: small seed mass, short juvenile periods, and short intervals between seed crops.[8] Not only do these three traits pick out invasive from noninvasive cases reliably. They also, as a result, successfully predicted two future invasions: first, dispersal of pine seeds from plantations to natural and semi-natural habitats in the Southern Hemisphere, and second, that some native conifer forests in Sweden would be invaded by the North American pine *Pinus contorta*.[9]

How was this unusual success achieved? By Case-Worker. The example illustrates two aspects of that strategy (each also emphasized by Elliott-Graves). The first aspect is empirical theory development. The theory—in this case, the three key traits and the mechanisms behind why they are important—is based on statistical filleting by Rejmanek and Richardson of data collated from many other studies of pine species. This initial inference step is correlational. It is then supported by plausible mechanisms and by successful predictions. Unlike the Prisoner's Dilemma, the theory was not developed according to internal or abstract criteria, nor was it derived from a single, cross-contextual master model. Just as well, too, as pines' invasiveness was a surprise—the fresh data

[7] Rejmanek and Richardson (1996); Richardson and Rejmanek (2004). See also Elliott-Graves (2016, 388–90).
[8] Rejmanek and Richardson (1996, 1657).
[9] Elliott-Graves (2016, 389). The predictions are in Richardson and Rejmanek (2004, 327).

analysis was crucial to the discovery. Conifers are often thought of as less successful evolutionarily than woody angiosperms, yet pines turned out to be "as
invasive (or more so) as any angiosperm family comprising predominantly
woody taxa. Indeed, only a few completely or prevailingly herbaceous plant
families seem to be more invasive."[10]

The second aspect of Case-Worker exemplified here is localist theory application. The theory applies reliably across the salient range of environments for
all pine species—but only for them. It is not reliable for other species. Attempts
by Rejmanek and Richardson to extend the theory to invasive woody angiosperms, and even just to other conifers, soon led to predictive failures.[11] For
example, efficient animal dispersal can mean that even large-seed conifer species are invasive. Small seed size is no longer necessary. Outside of pine species,
the mechanisms that operated reliably for pine species sometimes break down,
and to explain the fate of invasions becomes again impossible without additional contextual work.

In sum, only a locally reliable model could be found and only by Case-
Worker theory development.

In ecology, "crypticity" is widespread: many relevant properties of species
and environments cannot be predicted ahead of time.[12] But blame nature for
that. And then, use methods appropriate to this reality. Philosophers of science should lionize those ingenious scientists who have wisely foregone the
Stability-Theorist strategy and who, in the case of pine species invasions, have
achieved success by pursuing Case-Worker instead.

5.2 Arguments from first principles

Turn now, in the next few sections, to why we should expect fragility to be
widespread. Begin with some arguments from first principles.

A variable's behavior can be unpredictable even when its underlying dynamics are not.[13] In systems that are mean-reverting, for example, the impact
of a given change in input varies depending on exactly when it occurs, eventually dropping to zero as the system reverts to the mean. So, while the underlying function is predictable, the relation between change of input variable
and size of initial impact on the output variable is not. Or suppose a system is

---

[10] Richardson and Rejmanek (2004, 325).
[11] Rejmanek and Richardson (1996, 1658–9); Richardson and Rejmanek (2004, 326–7).
[12] Jarić et al. (2019).
[13] Sugihara et al. (2012); May (1976, 1977).

subject to a tipping point or critical transition. In that case, while nudging it by a certain amount normally has only a small or zero impact, beyond a certain critical level the same increment of nudge may have a dramatically huge impact, as when we gradually increase the pressure on a pencil until it suddenly snaps. The relation between the degree of added nudge and change in output varies.

Although developed in ecology, these mathematical scenarios are perfectly general. Many systems from other domains have the same formal properties.

Are these kinds of relation, to use our terminology from Chapter 2, merely "stable-contextual" rather than fragile—that is, are they sensitive to background conditions but still predictable? If we know the stable underlying function, the objection runs, and if we know enough to calibrate it, then we can reliably predict the impact of interventions, after all. We can know, for example, how close to the mean a variable has already reverted, or whether a variable has reached a tipping point. But often, we do not know. We do not know, for example, exactly when a pencil will snap. So, although Stability-Theorist makes sense here for Laplace's Demon, it does not make sense for us.

Similar conclusions follow from a different analysis—complex adapted systems theory. Developed in medicine and public health, this argues that in many domains, multiple factors interact in dynamic and unpredictable ways. Causal relations are not stable.[14] Rather, the underlying causal structure is constantly shifting, as different interacting factors come and go. This means we should not test for whether an effect size is statistically significant once other variables have been controlled for because that presupposes there is a stable effect size to be found. Instead, contextual, rapid-cycle evaluation is preferred:

Does this intervention contribute, along with other factors, to a desirable outcome [in a given case]? Multiple interventions might each contribute to an overall beneficial effect through heterogeneous effects on disparate causal pathways, even though none would have a statistically significant impact on any predefined variable. To illuminate such influences, we need to apply research designs that foreground dynamic interactions and emergence. These include in-depth, mixed-method case studies (primary research) and narrative reviews (secondary research) that tease out interconnections and highlight generative causality across the system.[15]

---

[14] Rutter et al. (2017); Greenhalgh and Papoutsi (2018).
[15] Greenhalgh (2020).

In other words, we need the Case-Worker strategy.

A third line of analysis ends in a similar place. Brian Beckage and co-authors, working in the tradition of Stephen Wolfram, trace degrees of *computational irreducibility* across different kinds of systems.[16] If something is computationally irreducible then, by definition, it is not predictable by any simplified model or amenable to theoretical summary but instead must be investigated in detail, case by case. Stability-Theorist is ruled out.

Beckage et al. offer a three-way taxonomy, with escalating degrees of computational irreducibility. Roughly, *physical* systems become irreducible because of macro-level chaotic effects, *biological* systems become more irreducible because of Darwinian evolutionary effects in addition to chaotic ones, and finally, *human social* systems become yet more irreducible because there is added free will and sentience effects. The analysis claims to explain, perfectly generally, why relations in these domains are likely to hold unpredictably.

Should we accept Beckage et al.'s conclusions? They are not based directly on empirical study. Many relations even in the biological and social domains do hold predictably, as numerous examples illustrate.[17] But Beckage et al.'s analysis does give a reason why, in those domains, we should expect fragility—and permanently so, against the hopes of "one more heave."

Long ago, Michael Scriven argued that Newton-style stable laws are unlikely in social or psychological sciences.[18] Others, too, as we will see in Chapter 7, have argued that certain features of social science make fragility more likely.

5.3 Contrastive explanation

Contemporary theories of explanation, especially causal explanation, are predominantly *contrastive*.[19] We seek to explain not just why an event X occurred, but rather why X *rather than* Y occurred, or why a variable X took its actual value *rather than* another value. Explanation is sensitive to the precise specification of an explanandum. To take a fateful example, does the fine weather on 6 August 1945 explain why the atomic bomb was dropped on Hiroshima that day? Yes, if we mean why it was dropped on that day rather than the previous day, but no, if we mean why it was dropped in that year rather than the previous year.[20]

---

[16] Beckage et al. (2013).
[17] See also Ghomi (2022).
[18] Scriven (1956).
[19] Dretske (1972); van Fraassen (1980); Garfinkel (1981); Achinstein (1983); Woodward (2003); Northcott (2013b).
[20] The example is from Northcott (2013b).

Other work supports this contrastive focus. Elliott Sober argues that empirical testability in general requires a contrastive formulation of an explanandum.[21] Several authors have argued that causation itself, not just causal explanation, is contrastive.[22] A contrastive form is implicit in the contemporary Bayes net and causal modeling literature, and it is also endorsed by experimental practice, at least in quantitative cases.[23] It is consistent with the mainstream literature on probabilistic causation.[24] And causal explanations and assignments of causal responsibility in many sciences are implicitly committed to a contrastive scheme too—examples include arguments and measures in history, psychology, psychiatry, statistics, epidemiology, law, and computer science.[25]

Why does this matter? Because contrastive theories of explanation imply that, for any actual event, there are many associated explananda, and thus many different relations are potentially explanatory. It is easy to generate explananda that require either stable or fragile explanans relations, simply by appealing to a contrast that is implied by, respectively, a stable or fragile relation. Given that the number of potential contrasts is unlimited, in practice, this can always be done. So, for any given event, both stable and fragile relations will be explanatory, each with respect to different contrasts. This is the sense in which fragility is ubiquitous. (So too, therefore, is stability. In this way, two opposite properties—fragility and stability—may each be ubiquitous. But the emphasis of this chapter will be one-sided, on fragility.)

The same point may be made in other, familiar ways. There are many true *descriptions* of any target. Some descriptions pick out some causes or aspects of the target, others pick out others; some thereby highlight a stable explanans relation, others a fragile one. And usually, we have different *interests* in the same target; some interests highlight stable aspects of the target, others fragile aspects.

It is too simple to say that a target is characterized by stability or fragility in any absolute sense, let alone that a whole domain is. There are no such uniform patches. Rather, things are always contrast-relative, description-relative, and interest-relative.

Some examples illustrate. Why does a raindrop fall? Stably explained by gravity. Why does it fall to this exact spot on the grass rather than to another?

---

[21] Sober (1999).
[22] Maslen (2004); Schaffer (2005); Northcott (2008a).
[23] Woodward (2003); Pearl (2009); Spirtes et al. (2000).
[24] Hitchcock (1996).
[25] Northcott (2008b, 2013b); Kaiserman (2018).

Not explained by gravity, but instead only by local relations that are likely fragile.

Or consider earthquakes. These follow a power law: strong earthquakes occur less frequently than weak ones, and charting strength against frequency gives an exponential curve. This relation holds stably. So too, do other features of earthquakes, such as that they usually occur near geological fault lines, that they are usually followed by aftershocks, and that seismic waves spread through the Earth in certain ways. But what about *individual* earthquakes? These are notoriously hard to predict: even when an earthquake is thought likely soon in some region, to predict its precise time, place, and strength is still almost impossible, and to explain things after the fact is difficult too. The relations that speak to these fine-grained explananda are fragile, unlike the power laws that speak to aggregate data. Extensive local investigation is required.

Or consider wars. Again, there are well-established power laws that relate severity to frequency, but these do not predict *individual* wars well. There are many tendencies, correlates, and rules of thumb that do predict and explain individual wars—but only sometimes. As with invasive species, none works reliably.

Conversely, within this book's own case studies of fragility, one can find explanatory angles that bring stable relations to the fore. Why was a species invasion successful? That might require fragile relations to explain. But why did the invasion take several months rather than several hours? That is explained by stable relations that pertain to species' movements, reproductive cycles, and so on.

None of this shows that, in practice, *salient* contrasts or descriptions or interests implicate fragile relations frequently; for evidence of that, see the rest of this chapter, and book. But it does demonstrate that fragility is not some faraway oddity. Rather, it is always lurking not just around the corner, but right in front of us.

5.4 External validity and extrapolation

Problems with extrapolation, and with the closely related issue of external validity, are endemic in field sciences.[26] This is a truism. And these problems immediately bespeak fragility: a relation established in one situation cannot be relied on to hold in a new situation.

---

[26] External validity is interpreted in more than one way (Nagatsu and Favereau 2020). I will mean by it, roughly, that a finding in one context may be taken to hold true in a different context.

The need for extrapolation is ubiquitous in both science and policy, but it is difficult to know in advance when an extrapolation will work. To extrapolate a predictive model or causal relation to a new domain, we need detailed knowledge of this new domain, but if we have such knowledge already, then the extrapolation is redundant. Daniel Steel has christened this the "problem of the extrapolator's circle."[27] It implies fragility.

Several solutions have been suggested.[28] Simple induction from one domain to another is perhaps the most common "solution" in practice. The next most common is to rely on knowledge of mechanisms, which are presumed to operate stably across domains and thereby to license extrapolations. But mechanisms in field sciences are not always stable in the right way: they are typically not fully understood, are unshielded, interact with other mechanisms in intricate ways, and are highly sensitive to initial conditions. They are not like cars and toasters. The problem of the extrapolator's circle returns: to know how a mechanism will behave in a new domain, knowledge of that new domain is required.

The same conclusion follows from general considerations of evidence, at least in the causal case. All causal relations require support factors.[29] The justification for any extrapolation is then three-part.[30] We need good reason to believe that: (1) the relevant mechanism was at work in the old domain, (2) it will play the same causal role in the new domain, and (3) the mechanism's support factors will be present in the new domain. If we lack good reason to believe even one of these conditions, the justification fails. But it is impossible to justify the claims in conditions (2) and (3), and thus to justify an extrapolation, without extensive knowledge of the new domain. In which case, the problem of the extrapolator's circle has no solution.

This negative conclusion is widely accepted: the relations that underpin policy interventions or field trials in one context typically cannot be assumed to carry over to another context.[31] They are too fragile for that.

The only exceptions are "trumps." These are mechanisms or relations that are robust across all relevant environments, that is, whose operation is not interfered with by other causes and whose required support factors are

[27] Steel (2008).
[28] Khosrowi (2019b) analyzes leading proposals in depth. See also Howick et al. (2013, 278).
[29] Mackie's famous (1974) notion of an INUS condition captures this idea.
[30] Cartwright and Hardie (2012, 54).
[31] Howick et al. (2013); Cartwright (2013, 2019); Grüne-Yanoff (2016); Khosrowi (2019a, 2019b).

either minimal or very widespread. Further, these properties must be known in advance, so they must be either self-evident or obvious from background knowledge. If they are, then we may, after all, be confident of successful extrapolation even without detailed knowledge of a new domain. But trumps are rare in field sciences. In social science, external validity is often questionable.[32] Scientists' own practice implicitly concedes the point: when field operation of a mechanism really needs to be ensured, extensive contextual investigation is demanded, as in the design of the spectrum auctions.

Other things point the same way. Results from laboratory experiments are notoriously unreliable in the field. And they can be unreliable even in other laboratories: the so-called replication crisis in parts of psychology and other disciplines suggests that a relation might no longer hold even when an experimental condition is ostensibly kept the same. In other words, these relations are fragile. (If a relation is fragile, then replication should not always even be *expected*, although I do not pursue that separate line of thought here.) Even aside from the replication crisis, external validity has long been a serious challenge in psychology. This is one reason that experiments alone, divorced from ecological and longitudinal studies, are of limited value.[33]

Similar concerns are widespread in economics. Results often do not replicate, even when bad practices are eliminated.[34] In development economics, for example, treatment effects are typically heterogeneous, and unpredictably so—that is, they hold unreliably.[35] The success of interventions is especially sensitive to local politics, and external validity cannot be assumed. Instead, Case-Worker is needed: "assessments of external validity will always remain best conducted on a case-by-case basis."[36]

5.5 Examples

Notoriously, there are few exceptionless generalizations in field sciences. Is this because field relations are typically fragile? Perhaps. But the case is not proven, because lack of empirically accurate generalizations is also compatible with stability plus noise. And if there is a bedrock of stable relations, then science

[32] Levitt and List (2007); Reiss (2008, 92–6).
[33] Diener et al. (2022).
[34] Huntington-Klein et al. (2021).
[35] Kerwin (2021).
[36] Vivalt (2020, 3086).

could progress by identifying these and then filling in further sui generis factors as needed—a strategy advocated by Mill and by many others since.[37]

So, is there indeed a bedrock of stable relations, available to be exploited? Is nature kind? What matters methodologically, remember, is that stable relations are relevant and accessible. If our goal is to explain species invasions, for example, does it help if relations between subatomic particles are stable? Obviously not. Rather, the stability that matters is that of the mechanisms that explain species invasions: islands rather than mainlands, temperate climates rather than tropical ones, plant–soil feedback, and the rest—are any of *these* stable?

A key indicator of stability, as we saw in Chapter 4, is accurate interventions: if relations hold reliably, we should be able to *do* things with them reliably. Can we? Not in our case studies. Hypothetical interventions, if we could make them, would not be reliable. Switching from an island to a mainland, or from a temperate to a tropical climate, for example, would not reliably block the success of a species invasion. Facing each other for long periods did not reliably lead soldiers to truces in the World War One trenches. Interventions based on della Porta's explanations of Italian and German political violence would not work reliably beyond her Italy and Germany cases, as she herself emphasizes. And no auction model gave reliable advice about interventions when designing the spectrum auction. Recall, for example, from Chapter 4 (italics added):

In some models an open auction design (i.e., no sealed bids) reduces the "winner's curse" effect, by which the winner of an auction overpays because they are the bidder who most overestimates the prize's true value. But this may no longer be true when the value of a particular spectrum license depends on what other licenses a bidder wins ... In the experiments, more features were added specifically to defeat the winner's curse, namely, flexible minimum increments that made it more attractive to bid when activity was low, and also activity rules that required bidders to submit bids on pain of losing their eligibility. *Designers could not trust that the capacity of an open auction to defeat the winner's curse was stable enough to be relied on.*

Is fragility common, beyond our case studies? I believe so. Or at least, I think this claim is plausible, even though no probative measure of fragility's frequency across science seems feasible. To support the claim that fragility is

---

[37] Mill (1843); Cartwright (1989); Kincaid (1996).

common, there is value in surveying many examples relatively quickly, and in the rest of this section I do that. These brief reviews complement the rest of this chapter's more conceptual arguments.

Begin by following on from the Prisoner's Dilemma example. A closely related notion is the "tragedy of the commons"—roughly, the idea that if not prevented by regulation, agents will deplete a resource through their uncoordinated actions, contrary to the common good of all. Although it dates from the nineteenth century, the idea was made famous by a 1968 article by Garrett Hardin.[38] It has been applied to many cases of conservation. Yet as Nobel-winning economist Elinor Ostrom and others have shown, in cases of animal conservation, the tragedy of the commons is frequently *not* observed.[39] We require instead a toolbox of more sophisticated models, where local peculiarities of geography, history, and politics are integrated to give a different explanation each time. Narrowly pursuing just one master model such as the tragedy of the commons is exactly how not to make progress, much as it was not with the Prisoner's Dilemma. Ostrom's work has been criticized because, although she too starts with the Prisoner's Dilemma, it lacks a unified theoretical treatment. But why think of the tragedy of the commons, or the Prisoner's Dilemma, as the baseline, and deviations as of merely local interest? That framing loads the dice. Given the frequent actual deviations from both the tragedy of the commons and the Prisoner's Dilemma, it is better to see these just as two more theories in our toolbox, that like other theories sometimes apply and sometimes do not. To privilege these master models over more local analyses can be motivated only by an aspiration for universalism, but as I discuss in Chapter 8, such universalism is a wrong turning when explanans relations are fragile—and in animal conservation cases, it turns out, fragile they usually are.

Turn next to two examples from natural science. The first is the story of a tragedy.[40] In 1963, a huge landslide in the Vajont valley in north-east Italy caused lake water to overflow a dam and flood a downstream village. 2,000 people were killed. The dam that led to this disaster—by creating the lake—had been built only four years earlier. Why was the disaster not foreseen and prevented? Engineers knew that a landslide could cause flooding, but they were confident that no landslide would happen because they thought the limestone slopes around the lake were secure. Initially, no back-up observational study was carried out to confirm that. Why not? Because the surfaces of the slopes were

---

[38] Hardin (1968).
[39] Nijhuis (2021).
[40] Barrotta and Montuschi (2018). See also Cartwright et al. (2022, 119–26).

unflawed and because the following relation was assumed to be stable: "limestone is honest because it reveals its flaws on its surface."[41] It follows that limestone without flaws does not collapse in a dangerous landslide. But the "honest limestone" relation turned out to be fragile, not stable: it could not be relied on. Mistaking it for a stable relation led to neglect of supplementary investigation and thus to error—and in this case, to tragedy. In fact, after various alarms were raised—by locals, by some smaller landslides, and by some but not all geologists—supplementary investigations were begun. But alas, too late.

Here is a second claimed example of fragility from natural science. John Dupré has long advocated what he calls a process view of biology.[42] According to this view, roughly, biological mechanisms and dispositions do not operate reliably across environments. We should not study viruses, for example, in isolation, but instead should take the dynamic and context-dependent nature of viruses seriously and study how their dispositions interact with the environment rather than think those dispositions operate invariantly.[43] In other words, Case-Worker not Stability-Theorist. And the same lesson applies, Dupré argues, to living things generally.

Turn to an example of fragility from engineering: Formula One race cars.[44] Designers must maximize a car's speed and reliability while constrained by regulations about weight, tires, engine size, fuel capacity, and many other details. These regulations change with each championship season. There is a lot of relevant theory to know, and senior designers are highly trained engineers, yet knowledge of stable theoretical relations is not enough. All teams have huge testing programs, analogous to the experimental testbeds of the spectrum auction: new chassis designs are tested extensively in wind tunnels, while in-house drivers take new models for timed laps on private tracks. These wind tunnels and practice laps are essential to success—success correlates with development budgets rather than with theoretical knowledge. The key point for us is that the knowledge accumulated during development and testing is fragile. Will a given tweak improve the car, for example? The answer is highly contextual and needs to be re-established anew with each change to the rest of the car, in the manner of Case-Worker rather than Stability-Theorist. No one knows how different tweaks will combine except via trial and error. This is reflected by limited exportability: the success of one season's car does not guarantee success under

---

[41] The sentence is from a letter written by the dam's chief engineer, quoted in Barrotta and Montuschi (2018, 21).
[42] Dupré (2012, 2013, 2021).
[43] Dupré and Guttinger (2016).
[44] Alexandrova and Northcott (2009); Youson (2020).

new regulations in the next season, as the sporting record has demonstrated many times.

Turn next to medicine. Begin with chemotherapy, about which I will go into a little more detail.[45] The impact of drugs can be unpredictable. The chemical cisplatin was first tested as a therapy for some cancers in the 1960s, but positive effects proved sadly small and short-lived. Then in 1974, it was combined with two other chemicals, bleomycin and vinblastine, in a combination treatment for advanced testicular cancer. At the time, the survival rate for patients with this condition was under 5%, but the new treatment proved a sensation: almost every patient who received it improved enormously and, amazingly, a majority was fully cured, that is, never did relapse. This success was a surprise. Vast numbers of chemicals have been trawled by cancer researchers for signs of chemotherapeutic potential, and no one especially suspected that cisplatin would be one of the few jackpots. There was a second surprise too, namely the contrast between the effect of cisplatin alone and the effect of cisplatin in combination with the other two chemicals.

Research in chemotherapy typically takes one of two approaches:

(1) Black-box, trial-and-error trawling of chemicals, with little or no attempt to predict winners from the underlying biology. Cisplatin is a poster child for this approach.
(2) Targeted attempts based on an understanding of a cancer's underlying biology.

Is fragility characteristic only of (1)? Examination of targeted cases suggests not.

A kinase is a protein that tags other proteins in a cell with a phosphate group, leading to a cascade of signals. Chronic myeloid leukemia (CML) is associated with an aberrant gene, named *Bcr-abl*, which codes for a particular form of kinase that escapes the usual tight cellular regulation and runs haywire. It was therefore thought that a kinase *inhibitor* would be a potential therapy for this cancer. Geometrically, kinase molecules have deep pockets, and their action can be blocked by another molecule that is just the right shape to "fill" that pocket, much as a key fits a lock. The task was to find a chemical that inhibited specifically the kinase implicated in CML. One such chemical was eventually synthesized, and the treatment developed from it, under the commercial name

---

[45] For details and references about the next few paragraphs, see Mukherjee (2010). For the philosophy of science of cancer generally, see Plutynski (2018).

Gleevec, has proven highly effective against CML. It is a poster child for the targeted approach to chemotherapy.

The rough lock-and-key story above is as far as theory takes us, though. Throughout the 1980s, researchers at Swiss pharmaceutical firm Ciba-Geigy synthesized and tested literally millions of chemicals before they found one that was effective against kinases and simple enough. Many thousands of variants of this chemical in turn were then tested for leukemia specificity, nontoxicity, solubility, and other features. That is, exactly which kinase inhibitor would work (let alone in what dosages or with what implementation schedules) could not be determined except via painstaking trial and error. No one knew why Gleevec proved to be effective, only that it was. The full mechanistic story is still unknown today.

In sum, there is stability—Gleevec cures reliably, as does the cisplatin combination treatment. But there is also fragility—the impact of a kinase inhibitor could not be successfully predicted until it was tried, and neither could cisplatin's.

Chemotherapy is far from the only case of fragility in medicine. In nutrition science, few results are extrapolated easily. Typically, the impact of a nutrient in one product in one diet cannot be relied on to hold in a second product in a second diet. As well as diet, other aspects must be investigated too: dosage, the food's intended use and its form and frequency of intake, and aspects of a study group's lifestyle. Surprises are common.[46]

Arguably, fragility is widespread in medicine generally.[47] The same is true of epidemiology. Many relations central to the COVID-19 pandemic were fragile, as I discuss in Chapter 10. And it is not just COVID. The effect of HIV-AIDS therapies, for example, also varies unpredictably across contexts.[48]

Turn next to an example from cognitive science. Do theories from extended and distributed cognition apply predictably? Field studies ("cognitive ecology") suggest that they do not. One well-known example is the navigation system of a US naval frigate, which involves many sailors and instruments, many of which operate stably. But according to Edwin Hutchins, the complex interactions of different components mean that, with respect to the overall navigation task, actions have unpredictable effects, that is, they are fragile. The most productive methodology is qualitative, contextual inquiry, akin to Case-Worker.[49]

---

[46] Bengoetxea and Todt (2021); Neale and Tapsell (2019); Jacobs and Tapsell (2013).
[47] Solomon (2015); Stegenga (2018).
[48] Seckinelgin (2017).
[49] Hutchins (1995). I thank John Sutton for bringing this example to my attention.

Turn now to social sciences. Relations that hold reliably across contexts are famously rare in social science, but here are some of the few suggested exceptions:

- Law of Demand: the price of a good is inversely related to the quantity demanded.
- Okun's Law: change in unemployment is inversely related to the growth rate of output.
- Say's Law: supply creates its own demand.
- Iron Law of Wages: in the long run, real wages move toward subsistence level.
- Duverger's Law: a plurality voting rule favors a two-party system, proportional representation favors a multiparty system, and plurality with a second-round run-off favors a multiparty system.
- Iron Law of Oligarchy: all organizations move toward rule by a small elite, even those that are initially organized democratically.
- Malthus's Law of Population: population growth is exponential.

Julian Reiss details how much work is needed to establish predictability even in these favorable cases: can we find out all the conditions required for these relations to hold, and so make them predictable?[50] Sometimes, we can. Reiss argues that the Law of Demand is an example. But the wider pattern is that often we cannot, and then the relation is fragile—and this is true, remember, even of what are especially promising candidates for stability. The takeaway: in social science, fragility is likely commonplace.

This general conclusion is supported by many examples. In criminology, key relations such as that between probation policy and recidivism, or between catchment areas for spousal abuse shelters and degree of take-up, do not hold reliably.[51] Extrapolation is perilous: "it may be impossible to predict changes resulting from interventions without actually intervening."[52]

What determines election results? What is the relation, for example, between economic growth in the two quarters before polling day and an incumbent's probability of winning? The answer is that this relation varies, and it varies unpredictably. That is, it is fragile. And as I discuss in Chapter 9, the same is true

---

[50] Reiss (2017).
[51] Berk and Freedman (2003).
[52] Berk and Freedman (2003, 251).

for other causes of election results, and for related matters too, such as what factors influence turnout, or which campaign tactics are most effective.[53]

In economics, Julian Reiss argues that scientists' own practice implies that, in this book's terminology, causes are fragile. For example, a famous study showed that, in one circumstance, raising the minimum wage increases employment.[54] But in other circumstances, it does not: say, when the minimum wage is already high, when it is raised by a large amount, or when economic conditions are different. In response, crucially, rather than search for countervailing causes that outweigh the original employment-increasing one, researchers just assume that the original employment-increasing cause no longer obtains. In other words, they assume it holds unreliably. Such fragility, Reiss argues, is typical in economics.[55] I return to economics in Chapter 8.

What determines economic growth? To predict and explain gross domestic product has proved difficult (see Chapter 9).[56] Given the complexity of what determines a country's gross domestic product, likely no one forecasting method captures all of the generating processes, but even if one did, likely those generating processes themselves are fragile, for several reasons. First, the economy is an *open* system. In other words, it is continuously impacted by unpredictable noneconomic factors, such as election results, that inevitably do not appear in economic forecasting models, and that disrupt the operation of the relations relied upon by those models. Second, the economy is likely a *chaotic* system, which makes explanans relations less likely to be stable. Third, one recent argument holds that any stable causal relations in macroeconomics must incorporate agent expectations, but because such expectations are unobservable, any causal relations that *we* can establish will not be stable.[57]

What of forecasting in general? As I discuss in Chapter 3, success requires, in effect, a Case-Worker strategy. Forecasts that rely solely on a putatively stable theory are usually accurate only within highly shielded settings such as laboratories or artifacts. Out in the wild, they go wrong. The lesson of Philip Tetlock's long-running series of forecasting tournaments is that the approach of the best forecasters is piecemeal and contextual.[58] They suck in expertise and information from many sources, and they use general theories only as a useful starting

---

[53] Northcott (2015).
[54] Card and Krueger (1994).
[55] Reiss (2008, 173–6). See also Reiss (2012).
[56] Betz (2006); Northcott (2020).
[57] Henschen (2018).
[58] Tetlock (2005); Tetlock and Gardner (2015).

point, certainly not to be presumed to hold without further vindication—the opposite of Stability-Theorist.

Similar remarks apply to prediction markets. These have a sustained record of relative forecasting success.[59] The mechanisms behind this success again emphasize disaggregated, contextual knowledge, integrating multiple lines of evidence rather than relying on a single theory. Traders are forced to be guided by what strategy works, and that strategy is Case-Worker not Stability-Theorist.

Lastly, other cases of successful forecasting, such as by sports bettors, are likewise achieved by emphasizing contextual rather than theoretical knowledge. And experience of forecasting with big data methods supports a similar conclusion (see Chapter 9).

---

[59] Tziralis and Tatsiopoulos (2007); Horn et al. (2014).

6

Fragility and philosophy of science

6.1 Laboratory versus field sciences

In the remaining chapters of the book, I move from fragility to its applications. I will discuss, from a fragility perspective, a range of issues in science and philosophy of science that have received many methodological commentaries, in the hope that these commentaries can be connected and thereby deepened.

I begin, in this chapter, with several issues within philosophy of science. To start, consider some familiar and effective solutions to the problem of noise: laboratory experiments, field trials, and many statistical methods. Fragility bears on all of these. I then discuss qualitative methods. Lastly, I move on to the classic (within philosophy of science) topics of scientific realism and scientific explanation. I defer the discussion of reflexivity, economics, big data, and epidemiology to later chapters.

Laboratory methods are well suited to stable relations. Controlled, shielded experiments avoid underdetermination and noise. Inference is made possible, and the relations inferred may then, because stable, be applied outside the laboratory with confidence. In the best cases, laboratory experiments license interventions throughout the world—a remarkable bounty.

When relations are fragile, it is a different story. Now, success in the laboratory does not extrapolate reliably to the field, so fresh empirical warrant is needed for each application anew—a much weaker bounty. It is no coincidence that theories in laboratory sciences tend to focus on stable relations instead. Perhaps it is also no coincidence that philosophy of physics has largely ignored the methodological issues discussed in this book: many branches of physics deal with relations that are stable.

Past philosophy of science has often seen the contrast between laboratory and field sciences differently. Stable relations have been thought to characterize both kinds of science alike, progress made by identifying these relations, and individual cases tackled—Mill-style—by adjusting contextually for disturbing

Science for a Fragile World. Robert Northcott, Oxford University Press. © Robert Northcott 2025.
DOI: 10.1093/9780191944352.003.0006

causes.[1] What, then, on this view, is the difference? Noise. In laboratory sciences, it is easier to shield from noise, and so the epistemological circumstances are more propitious. But there is not thought to be any systematic difference methodologically—it is Stability-Theorist either way.

I disagree. Noise is not the crucial factor; fragility is. In laboratory sciences, relations tend to be stable, and this spontaneous sorting makes good sense. But it is not perfect—sometimes, laboratory methods are applied when relations are fragile. While even in these cases shielded experiments can develop the theoretical toolbox, an obvious limitation lurks: laboratory tests of a fragile relation are not a reliable guide to how that relation will behave in the field.

Recall the Prisoner's Dilemma.[2] A lot of the Prisoner's Dilemma literature is theoretical development, while very little of it is close empirical analysis of field phenomena. But a third category, although hard to quantify precisely, might form the largest portion. This third category is psychology experiments, especially simulations in the laboratory of the Prisoner's Dilemma or closely related situations. For example, how much more likely is cooperation (in a laboratory) if we use labeling cues, if we vary payoffs asymmetrically, if players have a prior friendship, if players have an empathetic personality type, or if players expect cooperation from opponents?[3] Literally, thousands of articles are similar.

How effective are these experiments at establishing relations for field use? Unfortunately, not very. Relations that hold in the laboratory do not predict or explain reliably the behavior of banks, firms, consumers, and soldiers in the field. Steven Levitt and John List discuss this for cooperation specifically, and like everyone else, they conclude that the external validity of laboratory findings can rarely if ever be assumed.[4] Findings fail to extrapolate even to cases that one might think especially close to laboratory conditions and thus especially promising candidates for Prisoner's Dilemma analyses, such as highly rule-confined TV game shows.[5] In fact, the same is true even within laboratories themselves. Much of the literature details how the Prisoner's Dilemma's predictions break down in experiments, and how instead we need a richer account, sensitive to contextual factors.

In recent decades, the "new experimentalist" literature has greatly expanded our appreciation of the different epistemic roles that experiments play.[6] Testing

---

[1] Cartwright (1989); Kincaid (1996); and many others.
[2] The following three paragraphs are adapted from Northcott and Alexandrova (2015).
[3] See Northcott and Alexandrova (2015) for detailed references.
[4] Levitt and List (2007).
[5] van den Assem et al. (2012).
[6] Hacking (1983); Weber (2018); Franklin and Perovic (2021).

theory is only one of these roles. Others include refining and developing theory, learning from model systems or organisms, and investigating new phenomena to establish new evidence for future theory to explain. All of these other roles, though, are also best played *by laboratory experiments* only when relations are stable. Refinements of the Prisoner's Dilemma suggested by laboratory experiments, for example, are unlikely to be probative or fruitful compared to refinements suggested by applications in the field. When relations are fragile, we need case studies in the context of use or in contexts similar to it—and that is rarely a laboratory.

Laboratory experiments here are a wrong turning. They further a Stability-Theorist pattern of theory development that leaves us in explanatory poverty. An important part of the historical analyses that explain the World War One truces, for example, was qualitative work, such as interviews with veterans and narrative reconstructions of notable events. Laboratory experiments were no help with that.

Similar remarks apply to many areas of psychology. There are longstanding worries about, in effect, the fragility of relations discovered in the laboratory.[7] Participants may be crucially unrepresentative: usually, convenience samples are used, yet they are typically WEIRD (Western, Educated, Industrialized, Rich, and Democratic), and often just local undergraduates; experimental controls preclude learning the very patterns of contextual variation that could inform extrapolation; long-term effects can deviate from short-term ones; and there are frequent worries about construct validity—concepts such as "frustration" or "aggression" usually have no definitive operationalization, which creates a potential gap between the theoretical relation ostensibly being tested and what an experiment actually measures. These and other problems all tell against laboratory work, at least if that work is unsupported by other methods. The problems stem from fragility.

6.2 Statistical methods

There is a huge statistical methodology literature. How does fragility bear on it?

Statistical methods are tremendous tools for detecting signals from noise. But they overcome only noise, not fragility. Indeed, for most uses, they depend on *lack* of fragility: it needs to be assumed that the underlying generating process is constant throughout a sample.

---

[7] Diener et al. (2022); Möllenkamp et al. (2019).

Given stability, statistical methods are, functionally speaking, like laboratory experiments: they solve the noise problem, and therefore they solve the underdetermination problem. Inference is made possible. The added value of statistical methods is that they work even when laboratory experiments' solution to the noise problem—namely, shielding—is unavailable. The domains most helped by statistical methods are therefore those that feature both stability and noise, but where shielding is difficult. Examples are familiar and include many parts of medicine, psychology, biology, and social science. In these cases, stability often stems ultimately from the similarity of conspecifics: many aspects of physiology are similar across all humans, for example, as are many aspects of how humans interact with each other.

Sometimes in these domains, laboratory experiments are possible too. Then, stability enables a powerful combination of methods: laboratory experiments to establish a relation, followed by statistical work to establish the extent of that relation's operation in a noisy field environment.

Statistical methods can be useful in the laboratory too, not just in the field— when a stable relation is probabilistic. The relation might be probabilistic irreducibly, as perhaps in quantum mechanics. Or it might be that, in effect, we cannot shield off noise fully: sometimes in psychology, for example, the causes of human actions are too subtle to be predicted with certainty even in the laboratory.

There is a flip side to these successes, though: because they depend on stability, they disappear when relations are fragile. Consider political violence. As discussed in Chapter 4, the relations established by Donatella della Porta to explain the evolution of violent groups in postwar Italy and Germany do not hold stably. Any statistical regression across countries would fatally gloss over important differences between those countries, and so would miss many causal explanations. Only detailed, contextual work can establish those. The implicit aim of such statistical studies, namely, to confirm or discover a stable, wide-scope relation, is futile in this case.[8]

As noted in Chapter 2, the *strength* of a relation, over and above the relation itself, may be fragile too. So to speak, there is quantitative as well as qualitative fragility. Effect sizes in medicine, for example, notoriously vary—over time, within samples, and even within individuals (over time). Hence, the frequency of subgroup analyses. How exactly to define causal strength is an intricate matter, but on any plausible account, causal strength is highly sensitive to changes in background conditions, which makes it harder to predict

---

[8] Della Porta (1995, 14–20).

and, thus, more likely to be fragile.[9] The fragility of causal strengths immediately problematizes the interpretation of a regression coefficient, because such a coefficient is, in effect, a claim of a stable causal strength across a sample. True, a coefficient may be interpreted instead as merely an average across the sample, with no assumption of constancy (although such an interpretation does not seem to be common), but even then, such an average is interesting only if the underlying probability density function is stable. Otherwise, the average is purely an artifact of the individual sample, uninformative about anything wider.

Practitioners are well aware of these dangers, of course: nonstationarity is a classic concern in statistics, with many diagnostic tests. Still, the reliance of statistical methods on stability remains. The practically salient question becomes: how often are data-generating processes indeed stable? Perhaps not surprisingly, practitioners tend to be optimistic: "anyone who makes a living out of data analysis probably believes that heterogeneity is limited enough that the well-understood past can be informative about the future."[10] But as discussed in Chapter 5, there are reasons for caution.

A familiar consequence of fragility is lack of external validity. This is an Achilles heel even of randomized controlled trials, perhaps the most lauded of all statistical methods. An example illustrates this.[11] In 1985, a randomized controlled trial in Tennessee schools showed that smaller class sizes improved reading scores, especially for class sizes below 20 and for children who were disadvantaged. Encouraged by this, in 1996, California, faced by low reading scores in its public schools, implemented a program of reduced class sizes, at a cost of over a billion dollars per year. It proved an expensive mistake—there was no significant impact on reading scores. Subsequent investigation revealed two main reasons: first, not enough classroom space (the available extra rooms in schools were often small and unsuitable), and second, not enough teachers (12,000 new teachers were required for the extra classes, and it proved impossible to recruit that many qualified ones in the time available). Without these crucial support factors, whatever mechanism worked in Tennessee did not work, or no longer applied, in California. That is, the causal relation between the policy change and reading score outcomes did not have external validity. As usual with fragile relations, the only remedy was detailed knowledge of the new context, that is, of schools in California.

---

[9] Kaiserman (2018); Northcott (2013b).
[10] Angrist and Pischke (2010, 23).
[11] Bohrnstedt and Stecher (2002); Cartwright and Hardie (2012).

Randomized controlled trials secure causal inference within their sample by allowing us to ignore contextual factors. That is their strength. The price, though, is less external validity, because knowing the influence of contextual factors is an important guide to where else the result of a randomized controlled trial will hold. Other guides are available, fortunately. One is background theory—if such theory is known to apply, either through supplementary investigation or because the relevant relation in the theory is stable.[12] Another guide is other forms of causal investigation.[13] In good cases, we may then be confident of a result's external validity, as with confidence that the success of COVID-19 vaccines in trials would extrapolate to the general population. But the point is that a randomized controlled trial alone is not enough.

Statistical methods' vulnerability to fragility bears on several influential recent movements in science and philosophy. The first is Evidence-Based Policy. This has been taken up widely: the World Bank has opened an Open Knowledge Repository, the UK now has nine What Works centers, and in the United States, what counts as scientifically based research in education was written into law in 2002 when setting standards for the Department of Education's What Works Clearinghouse.[14] "What works" here means, for the most part, policy interventions that have performed well in randomized controlled trials and sometimes in observational studies. But the relations that underpin these policy interventions are often fragile, so they hold only unreliably—as California found with small class sizes. This uncertain external validity is in tension with the ethos behind What Works centers, which promises a reliable guide for interventions, that is, an a-contextual reading of "works."

The second influential movement is the recent "empirical turn" in economics, which I discuss in Chapter 7. Pioneered by Nobel laureate Joshua Angrist and others, this movement advocates a range of experimental, quasi-experimental, and other methods for causal inference.[15] It emphasizes investigating "a-theoretically"—rather than relations suggested by theory, instead test for relations that were previously untheorized. Many aspects of the empirical turn I heartily endorse. But if pursued too narrowly, it runs into some of the problems discussed above: its favored methods of causal inference work only when relations are stable within a sample, with little emphasis on ecological or longitudinal studies, let alone on more informal or qualitative

---

[12] Deaton and Cartwright (2018).
[13] Diener et al. (2022). Both this paper and Deaton and Cartwright (2018) also criticize randomized controlled trials on other grounds, unrelated to fragility.
[14] Cartwright (2020a, 273).
[15] Angrist and Pischke (2009, 2010).

methods. And while it is good to avoid misplaced statistical testing of fragile theoretical relations, the neglect of wider theory does handicap assessment of external validity.[16] Angrist's "credibility revolution" is a big advance on what came before. But it cannot be a complete guide to policy interventions, just as What Works centers cannot, unless its methods are suitable not just for when relations are stable but also for when they are fragile.

Many of the same issues recur with a third movement: the causal modeling literature. Roughly, that literature defines causation in terms of directed acyclic graphs built around structural equations, where those equations represent the dependency relations presumed to underlie a particular situation.[17] This apparatus has a similar function to statistical methods: it enables inference from noisy data. It does so via sophisticated algorithms for inferring causal relations from patterns of probabilistic dependence, which improve on standard statistical methods by leveraging an explicit theory of causation. This is an important advance. As with statistical methods, though, causal modeling relies on stability: its algorithms must assume that the probabilistic dependency data are generated by a stable process, represented by the structural equations. Without that stability, the algorithms cannot work.[18] As with statistics, practitioners are well aware of this limitation of scope, but the fact of it remains.

A theme of this section is that statistical and causal modeling methods need the underlying data-generating process to be stable. It turns out that the same is true of big data methods, as I discuss in Chapter 9.

6.3 Qualitative methods

By qualitative methods, I have in mind a range of techniques, such as causal process tracing, case studies, small-n causal inference, narrative inquiry, "grounded theory," questionnaires and surveys, and ethnographic observation.[19] All of these can explain and predict single cases without assuming the stability of wide-scope relations, and all of them are compatible with a toolbox view of theory. So, I endorse qualitative methods.

A classic objection to qualitative methods is that single cases can be unrepresentative and therefore useless for establishing type-level relations, and so

---

[16] Leamer (2010).
[17] Pearl (2009); Spirtes et al. (2000).
[18] Cartwright (1999, 105–35) explores this limitation in detail.
[19] Brady and Collier (2010) and Rubin (2021) give introductions to, and defenses of, qualitative methods. See also any social methods textbook, such as Bryman (2016).

we are worse off than with quantitative procedures. But qualitative methods may also leave us *better* off. A mechanism might be directly observable rather than having to be inferred from statistics, which removes some worries about confounders and research design, and allows us to infer that a correlation is causal.

Qualitative methods may leave us better off in other ways, too. Ethnographic work can alert us to caveats and nuances that likely hold in only a fragile way, which adds to our toolbox of theories.[20] It can deepen causal knowledge from opinion surveys, by bringing out respondents' implicit commitments and why those commitments are held. And it can bring out how subjects themselves frame situations and options and how those framings evolve over time, thereby informing how abstract models are to be operationalized, and thus enabling experiments to run well. Some field experiments in development economics, for example, in effect employ an ethnographer on the ground.

With fragility, qualitative methods do not just provide supplementary support for quantitative inference, nor do they just help flesh out what mechanisms underlie quantitative inferences that have already been made; rather, they are also necessary for any quantitative inference in the first place. Suppose we use a randomized controlled trial to test for a causal relation. If that relation is fragile, then it cannot be assumed to hold throughout the trial's sample, so we must confirm that the relation's support factors hold throughout the sample, which in turn typically requires qualitative local work. In this way, Nancy Cartwright argues, even the "rigorous" method of randomized controlled trials rests on the back of the "unrigorous" methods used to confirm the auxiliary support factors (although, Cartwright also argues, these supposedly unrigorous methods are in fact perfectly defensible in the right conditions).[21]

Marcel Boumans comes to the same conclusion: any field measurement requires expert judgment and that judgment is often informal and qualitative.[22] Staying purely quantitative is a pipedream.

The same point is more familiar when applied to external, rather than internal, validity. To know whether an intervention will extrapolate to a new context, we need to know whether support factors will hold in that new context, which in turn often requires local, qualitative methods. Meta-analyses are little help here. Instead, we need field guides to where support factors or derailers

---

[20] Herzog and Zacka (2019).
[21] Cartwright (2021).
[22] Boumans (2015).

are likely to arise, and qualitative methods are likely needed to develop such guides.

All of these points are brought home by example. In the 1990s, the Moving to Opportunity economic experiment randomly distributed housing vouchers to poor families in Boston, offering them the chance to move to richer neighborhoods.[23] Before running a quantitative analysis, researchers conducted in-depth interviews, expecting families to be concerned primarily with quality of schools, as in earlier housing schemes. But things had changed. To the researchers' surprise, the primary driver for families now was fear of crime and the impact of that fear on health and on parents' desire to take jobs. The strength of the relation between the quality of schools and desire to move turned out to be fragile.

The interviews not only persuaded researchers to change the variables used in the subsequent quantitative analysis; they also enabled them to avoid a misinterpretation of that analysis' results. Black families in the scheme tended to move further away from the center of Boston than Hispanic families did. The qualitative investigation turned up, among other things, that this pattern was not (as the researchers had thought) because Hispanic families feared moving to areas with few Spanish speakers. Rather, it was caused by different advice from individual counselors assigned to the two groups.

This case is also an example where qualitative methods discover a mechanism that quantitative methods alone would likely have missed. Such knowledge of mechanisms helps with external validity. Here, will the causal relation between relocation and decreased stress for mothers continue to hold in other housing voucher schemes? One indicator to track, established by qualitative methods, is fear of crime.

The endorsement here of qualitative methods, unlike some other endorsements of them, is not committed to interpretivism about social science, which prioritizes "verstehen" understanding rather than third-person causal explanation.[24] Two notes here. First, contemporary empirical social science already meets some interpretivist concerns. The search is not for universal laws but for causal effects that are understood to be local; key concepts, such as "democracy," are debated extensively; all are keenly aware that actors' intentional concerns might disrupt statistical assumptions; and survey research rarely assumes simplistic fixed preferences in a single conceptual framework.[25] Second,

---

[23] Kling et al. (2004); Katz et al. (2001); Gennetian et al. (2012). I thank Cristian Larroulet-Philippi for bringing this example to my attention.

[24] Geertz (1973); Taylor (1971); Winch (1958).

[25] Lawler and Waldner (2023).

a fragility perspective also meets some interpretivist concerns, by endorsing qualitative inquiry in social science (and in natural science too). But it does so while embracing naturalist goals. When a relation is fragile, qualitative methods are endorsed *because* they are the best route to third-person causal explanations and accurate predictions, not because the latter are rejected as suitable goals.

A caution, finally: none of this implies we should reject the quantitative. Support for qualitative methods has often gone hand in hand with hostility to quantitative ones, and vice versa: "a tale of two cultures."[26] But both approaches are valuable, as the flourishing of mixed methods research illustrates. And quantitative, statistical analysis is often highly desirable when relations are stable—indeed, it can be essential. To assess the impact of an anti-drink-driving campaign, for example, accident statistics will likely be far more convincing evidence than a few individual case histories. Stability is ubiquitous (in the sense explained in Chapter 5). So too, therefore, is a potential role for quantitative methods. True, when relations are *not* stable, many statistical methods are no longer suitable, but even then, that does not tell against other forms of quantitative analysis. Donatella della Porta, for example, rightly eschewed a statistical study of political violence across countries but, along with much qualitative work, she also gathered a lot of useful within-country quantitative evidence too (see Chapter 4).[27]

6.4 Scientific realism

Turn now to the venerable scientific realism debate.[28] I will be concerned here with the epistemological branch of that debate, that is, with whether it is justified to claim that scientific theories give us the truth. It is this epistemological aspect that fragility bears on.

If a relation is fragile, then regardless of that relation's success elsewhere, without fresh empirical confirmation we have no warrant that a theory or model built around that relation truly describes a target phenomenon. If empirical warrant must be local, then warrant for realism about wide-scope models or theories is precluded. We may be realist not about scientific theories

---

[26] Goertz and Mahoney (2012), King et al. (1994), and Brady and Collier (2010) are two other influential salvoes in the debate.
[27] Della Porta (1995).
[28] Chakravartty (2017) surveys the huge literature.

but only about individual instantiations of them, or only about models or explanations that are narrow scope. In a fragile world, realism is local.

The classic perspective returns when relations are stable. Now, empirical warrant in one context does carry over to other contexts, and realism about wide-scope models or theories is potentially supportable again. Philosophical attention naturally returns to the question of whether empirical success is sufficient for realism.

What of instrumentalism? According to instrumentalism, roughly, theories are useful instruments for making predictions, but we are not warranted to claim that theories are true even when their predictions are accurate. It turns out that instrumentalism about theories also requires stability, just as realism about them does. To see why, consider an example.

Weather forecasting has become much better.[29] Hurricane paths are predicted more accurately and more in advance, and temperature and rainfall predictions are more accurate too. Overall, a few years ago, the accuracy of seven-day forecasts had become equal to that of three-day forecasts 20 years earlier. And today, the UK Meteorological Office estimates that four-day forecasts are as accurate as one-day forecasts 30 years ago.[30]

At the heart of weather forecasting models are equations of fluid dynamics that have been known for centuries. These equations govern (it is assumed) the fiendishly complex movements of air in the atmosphere, and how those are impacted by temperature, pressure, the Earth's rotation, the cycle of night and day, and so on. But the equations alone cannot generate accurate weather forecasts. Many ad hoc additions are required to accommodate factors such as mountains, clouds, or the coupling of air movements and ocean currents. Crucially, the exact form of these additions is underdetermined by fundamental theory, and indeed sometimes they contradict it.

Consider mountains, for example. They are well known to influence atmospheric circulation and to have large local effects on airflow and rainfall. In the early 2000s, it was explored whether introducing a term for the effect of mountains improved the forecasting model. To achieve that, it turned out researchers had to move beyond theory:

[One version of the model] included a "cut-off" or "effective" mountain height in the computation of gravity wave drag from the SSO scheme [i.e., the

---

[29] The following paragraphs are adapted from Northcott (2017, 2020), which contain details and references.
[30] Mance and Sheppard (2023).

scheme to represent mountains]. The more physically realistic cut-off mountain height resulted in a decrease in gravity wave drag (GWD), reducing the excessive deceleration of flow over the Himalayas and Rocky Mountains . . . . However, climate runs showed an increase in the positive zonal wind bias over winter northern hemisphere mid-latitudes, suggesting that the reduction in GWD had been excessive. This problem was solved [in the next version of the model] by doubling the "cut-off" mountain height and thereby increasing the amplitude of the gravity waves "generated" by the SSO scheme by a factor of two.[31]

That is, the *un*realistic formulation was the one eventually adopted, for brute instrumentalist reasons—it generated more accurate forecasts. Predictive fit trumped physical understanding.

For most of the twentieth century, weather forecasting models were interpreted realistically. They spoke of weather fronts, as when "a cold front from the west makes rain likely tomorrow," and they were explanatorily transparent. But more recently, progress has required an instrumentalist turn. Modern computers enable forecasters to run enormously complex simulations, testing and back-testing different models against the vast amount of data now available— running to tens of millions of observations per day. To be sure, models still start from a realistic ontology of seas, mountains, and so on, but they depart from that base. As with the mountains example, choice of model is driven purely by predictive success, in part for commercial reasons.

Here is the key point for us: to develop the weather forecasting model, we must assume stability; otherwise, testing against the weather data from yesterday would be an unreliable guide to forecasting that data for tomorrow. In return, the progress in forecasting accuracy vindicates this assumption of stability. To refine any theory enough to make it useful for prediction, that theory must hold stably over many instances.

To be a useful instrument implies some systematic usefulness across cases. In contrast, a narrow-scope model that predicts or retrodicts just one case correctly but that is known to be false is not normally termed a useful instrument. Rather, it is just termed "false" or a false explanation. As just noted, an instrument that is systematically useful across cases, like the weather forecasting models, can be developed only if based on relations that are stable. No stability, no instrument. So, no stability, no instrumentalism.

---

[31] Jung et al. (2010, 9). The page reference is to the ECMWF Technical Report.

Thus, the classic realist–instrumentalism debate applies only to half of science, so to speak—the stable half.

What of the fragile half? Fragility, recall, implies that empirical warrant, and thus potential warrant for realism, is only local. Should we be realist even locally? A successful intervention is certainly good evidence for a causal relation, but the mechanism behind that relation may still be underdetermined. With respect to underlying mechanisms, or to other unobservables, I do not see that fragility favors either realism or antirealism.

A recent strand in the scientific realism literature emphasizes "local realism" in a different sense.[32] It points out, roughly, that the arguments for and against realism vary greatly in different cases, reflecting the huge diversity of methods and evidence across science. Any universal recipe is too abstract to persuade. This localist strand is more open to realism about things other than wide-scope theories, because exactly what element of scientific work "latches onto reality" is conceded to vary greatly from case to case. Perhaps it is only more traditional, theory-centered versions of realism that apply poorly to a fragile world. (The argument above against instrumentalism in a fragile world, still holds, though.)

Ian Hacking's famous slogan, "if you can spray them, then they're real," defends yet another kind of local realism, namely realism about entities and their properties.[33] What Hacking's exact argument is has been debated, but the fact of successful interventions is clearly central to it, so might it carry over here?[34] I do not think so. The properties that Hacking has in mind, such as an electron's charge, are conceived as holding stably, so his claim is different from ours.

These connections between fragility and scientific realism, finally, naturally bear on the notion of scientific *progress*. One traditional view sees progress as scientific theories getting closer to the truth: much of the literature on the verisimilitude, or "approximate truth," of theories, conceives of progress in this way.[35] In turn, many contributors to the approximate truth literature are motivated by a desire to buttress scientific realism, understood as realism about theories. If relations are stable, it is more attractive to identify progress with theories getting closer to the truth, albeit there are other dimensions to progress too, as we will see shortly. But if relations are fragile, progress may not be seen in such a theory-centric way. Conceived of causally, for example,

---

[32] Saatsi (2017); Asay (2019); Vickers (2022); Magnus and Callender (2004). I thank Peter Vickers for helpful discussion.
[33] Hacking (1983, 23).
[34] Miller (2016).
[35] Popper (1972); Oddie (1986); Niiniluoto (1987); Kuipers (1987); Miller (1994).

we might progress by achieving more and more accurate causal descriptions, as demonstrated by successful interventions. But such progress could only be problem by problem, context by context. (I think this causalist approach is, in fact, the only promising way to make sense of approximate truth, even when relations are stable.[36])

This departure from the view that progress consists in theories getting closer to the truth clearly dovetails with the toolbox view of theories, one consequence of which is that theories lose exclusivity in their domains: if no single theory, then progress is not just refinement of a single theory. Rather, progress is better seen as increasing the size of the toolbox and improving the tools within it.

Much recent philosophy of science has explored such broader senses of scientific progress. Perhaps progress is best seen as an increase in knowledge.[37] Or perhaps it is best seen neither semantically nor epistemically, but instead pragmatically. In the latter vein, some emphasize how theories establish new concepts rather than getting closer to the truth, in a kind of conceptual or heuristic progress.[38] Others emphasize the development of models: many sciences are model based rather than theory based, and models, unlike theories, may be conceived as getting better at defining and solving specific problems rather than as getting closer to the truth.[39] A similar perspective sees progress as adding to a toolbox of mechanisms.[40]

More generally, many have emphasized scientific work beyond formal theories or models at all, such as contextual empirical investigations, or the development of equipment, techniques, measures, and informal know-how.[41] One example is progress in historical sciences. This often consists in deepening narrow-scope narrative explanations rather than in elaborating new mechanisms or theory.[42] Another example is responses to the replication crisis, such as mandatory pre-registration of studies. This is a kind of progress in practices: progress understood methodologically or even culturally.

---

[36] Northcott (2013a).
[37] Bird (2007).
[38] Alexandrova (2008); Alexandrova and Northcott (2009); Price (2019).
[39] Shan (2019).
[40] Hedström and Ylikoski (2010).
[41] Cartwright et al. (2022); Alexandrova and Northcott (2009).
[42] Currie (2014); Dresow (2021).

6.5 Mechanisms and scientific explanation

A rising force in recent philosophy of science has been the "new mechanist" program.[43] How does a fragility perspective bear on it? In brief, it supports it, but with a caveat.

The mechanist program comes with an associated view of explanation: roughly, to explain is to identify a causal mechanism behind an explanandum. A mechanism explains *why* a certain surface relation holds. Knowing this underlying story not only explains, according to the mechanistic view, it is also efficient methodologically: we understand superficially different phenomena in a unified way, extrapolate relations to new contexts more reliably, and are guided to fruitful new lines of research. (Critical realists, too, see the discovery of underlying mechanisms as a—even *the*—central goal of social science.[44])

It can be practically crucial, for example, to know by which of several possible mechanisms a policy works.[45] Mechanisms also illuminate in other ways. Consider the democratic peace hypothesis, according to which, roughly, democracies never go to war against each other. What explains this empirical regularity? According to Cartwright and co-authors, it is a mistake to think of a single, canonical explanation.[46] Instead, drawing on much political science, they highlight four distinct, overlapping mechanisms behind the regularity. Each of these mechanisms applies often; none applies always. But that is the point. The reason why the regularity holds so well is that it is overdetermined: multiple individually sufficient mechanisms each hold most of the time, independently of each other. A mechanistic view enables us to see this.

Mechanisms connote stability, but they may still explain even when relations are fragile. Fragility implies merely that the Case-Worker strategy is required to establish that a mechanism applies in a case at hand. Mechanisms' definitional stability is only with respect to their internal structure; it does not cover how a mechanism's output is affected by a changing environment.[47] And it is widely accepted that mechanisms, at least once outside shielded laboratories or artifacts, sometimes do not apply or extrapolate reliably. They may even be one-offs.[48] This has been argued in detail for many domains, including social

[43] Craver and Tabery (2019).
[44] Bhaskar (1975).
[45] Ylikoski (2021); Grüne-Yanoff (2016). Cartwright's (1999) notion of a "nomological machine" can play a similar role.
[46] Cartwright et al. (2022, 194–219).
[47] Pemberton and Cartwright (2014), Illari and Williamson (2012), and Craver and Tabery (2019) survey different definitions of a mechanism.
[48] Cartwright et al. (2018); Strevens (2012).

science and biology[49] and medicine.[50] When mechanisms do not extrapolate reliably, the explanatory warrant they provide is narrow scope. Mechanisms function in the same way as middle-range theories; indeed, middle-range theories often just describe mechanisms.

I am sympathetic to the mechanist program. From the beginning, it has understood mechanisms to apply only locally, and their role is seen as akin to tools in a toolbox. All this is to the good. Yet a fragility perspective endorses the mechanistic view only with a caveat. Why? Because of a methodological worry.

The mechanistic view directs research effort to the discovery and development of mechanisms, but this focus becomes a wrong turning if mechanisms are not developed empirically. Our case studies illustrate. The huge literature developing the mechanism of the Prisoner's Dilemma did not pay off for explaining World War One truces. Similarly, further elaborating game-theoretical auction models was not the way to produce an effective spectrum auction; rather, contextual experiments were required. And to explain species invasions, it is better to investigate locally, drawing on relatively simple rules of thumb, than to elaborate those rules of thumb in some abstract, a-contextual way.

The above worry concerns not the mechanistic view of explanation itself, but rather a practice that that view sometimes leads to—development of mechanisms without close empirical refinement. This practice departs from Case-Worker.

Narrative explanations trace a particular trajectory through time and so are narrow scope. Not by coincidence, they are common in historical sciences, where relations are often fragile. Adrian Currie has argued that a common form of narrative explanation is ill-suited to the mechanistic view because its goal is to tease out a particular causal history rather than to understand that history in terms of mechanisms that apply more widely.[51] As noted earlier, progress then consists in deepening narrative explanations rather than in elaborating new mechanisms. This worry anticipates our worry above, I think. Again, the objection is not to the mechanistic view of explanation itself: many narrative explanations may be seen as mechanistic ones. Rather, the objection is to what, in practice, the mechanistic view can lead to, namely, the development of theory divorced from the empirical nuances of actual cases.

These worries do not apply when relations are stable. And much of the advocacy for the mechanistic view concerns domains where relations often are

[49] Steel (2008).
[50] Howick et al. (2013).
[51] Currie (2014).

stable and where theory development is admirably empirical, such as neuro-science and other parts of biology.[52]

Turn now to so-called how-possibly explanations.[53] Rather than explain targets from the actual world, such explanations speak only to the idealized world of a model. They may have heuristic value as conceptual explorations.[54] Some have claimed more: that if we understand explanation broadly, then how-possibly explanations should count as explanations proper.[55] A related view is that the explanatory claims of models may be established by means of robustness analysis, that is, by showing that a model's derivations still hold even when its assumptions are changed.[56] Like how-possibly explanations, such robustness analyses too can be useful.

But a danger lurks. In some domains, such as rational-choice theorizing in economics, the following slip occurs: by expanding the reference of "explanation" to include how-possibly explanations and robustness analyses, we close off investigations prematurely, by giving a pass to theories that lack empirical confirmation. When relations are fragile, this is a mistake. Settling for theories that lack empirical confirmation is not only misplaced; it also risks forgoing alternative, more productive, avenues. Think of the intricate work that was required to explain—to how-*actually* explain—World War One truces, invasions by pine species, and the evolution of politically violent groups. This work should be lionized. By settling for how-possibly explanations, we instead endorse its neglect.

This danger is ultimately the same as the danger from the mechanistic view, namely that we encourage a theory-centered methodology that is not empiricist enough. As in the mechanistic case, the danger diminishes when relations are stable—theories now often do lead to how-actually explanations, and how-possibly explanations are less likely to lead us astray.

A fragility perspective also tells against the unificationist view of explanation.[57] According to the unificationist view, roughly, scientific explanation consists in giving a unified account of different phenomena, as when Newtonian theory gave a unified account of motion on Earth and in the heavens. But if an explanans relation is fragile, warrant for explanations is

---

[52] Craver (2007); Bechtel and Richardson (2010); Glennan (2017).
[53] Grüne-Yanoff (2009); Aydinonat (2008); Forber (2010).
[54] Craver (2007, 116–8).
[55] Ylikoski and Aydinonat (2014).
[56] Kuorikoski et al. (2010).
[57] Kitcher (1989); Friedman (1974).

typically narrow scope and therefore ill-suited to unifying diverse phenomena. Yet these narrow-scope explanations are explanatory, nevertheless.

All that said, for the most part, the main arguments in this book are independent of the scientific explanation literature. They hold—or not—on any plausible account of explanation.

Similarly, the main arguments in this book are, for the most part, independent of the scientific modeling literature, too.[58] They do not turn on the details of what models are ontologically, nor on exactly how models relate to theories. Nor do they turn on exactly how models represent their targets—indeed, in heuristicist cases, such as the spectrum auctions, what matters is only whether an eventual causal hypothesis represents its target, not whether an initial model does.

---

[58] See Frigg and Hartmann (2020) for a survey of the modeling literature.

7

Fragility and reflexivity

7.1 Reflexivity versus fragility

Reflexivity is, roughly, when studying or theorizing about a target itself influences that target. In this chapter, I argue that what matters is not reflexivity but fragility.[1]

There are several definitions of reflexivity and of related (or, as sometimes used, synonymous) notions such as reactivity and performativity.[2] Ian Hacking made famous the notion of shifting or changeable kinds.[3] On his account, reflexivity means that a kind is potentially altered by feedback effects, with each alteration of the kind potentially inducing reactions in the members of the kind that in turn further alter it, and so on, indefinitely. Hacking had in mind human kinds, that is, kinds that concern humans. Reflexivity has also been defined, more simply, as when a kind shifts because of human activity. Or been defined, more generally, as when theorizing impacts on objects of study, with shifting kinds merely one side effect of that. There are other definitions, too. Below, I articulate reflexivity in terms of shifting kinds, but for our purposes, nothing important turns on that.

Reflexivity is clearly distinct from fragility. All definitions of reflexivity have in common that theorizing impacts in some way on the target of that theorizing. But there is no reason that this impact must be unpredictable, and thus no reason that it must be fragile, so we may have reflexivity without fragility. There are also clear cases the other way around, that is, of fragility without reflexivity.

To assess the relative methodological significance of reflexivity and fragility, I will work through different combinations of the two, using a 2×2 table (Table 7.1). For each example, I borrow from detailed case studies already carried out by others.

---

[1] Most of this chapter is adapted with permission from Northcott (2022b).
[2] Jiménez-Buedo (2021).
[3] Hacking (1995).

Science for a Fragile World. Robert Northcott, Oxford University Press. © Robert Northcott 2025.
DOI: 10.1093/9780191944352.003.0007

Table 7.1 Reflexivity versus fragility

	Reflexive kinds	Nonreflexive kinds
Fragile relations	Case-Worker strategy Schizophrenia Autism	Case-Worker strategy Species invasions World War One truces
Stable relations	Stability-Theorist strategy Domestic dogs Gender roles	Stability-Theorist strategy Newtonian theory Electric toothbrush

Begin in the bottom-left corner, with reflexivity but not fragility. The first example is domestic dogs.[4] The kind "dog" is reflexive: repeated rounds of breeding over perhaps 15,000 years have changed dogs dramatically, both morphologically and behaviorally. Traits including tameness, obedience, teachability, shepherding, hunting, and certain physical characteristics, have been selected for, and what were originally wild wolves became domestic dogs and then later the many different breeds of domestic dogs today. Human interaction with the kind changed it. There were many rounds of Hacking-style looping effects as new kinds themselves stimulated new behaviors by breeders, which in turn fed back to change the kinds once more. The key relation is that between breeding and the evolution of dog traits. This relation is stable: we can usually predict the impacts of breeding interventions, as dog breeders exploit all the time, and we can explain these impacts by general Darwinian theory. Stability-Theorist works well. This is despite reflexivity and because of stability.

The example of gender roles teaches the same lesson: optimal method tracks fragility, not reflexivity. Briefly, the kinds "masculine" and "feminine" are reflexive (according to influential views) because they are strongly influenced by how people conceive of them.[5] But many relations involving them are stable. That is, we have reflexivity but not fragility. Indeed, Ron Mallon argues that reflexivity has itself been the stabilizing force, pushing the kinds back into line, so to speak, if they show signs of changing.[6] Because of this stability, wide-scope theories—in this case, sociological theories of gender roles—can reliably predict the impacts of many interventions and can explain them. For example, theory posits that being perceived as male causes an individual to be subject to certain expectations about behavior and appearance, in turn causing

---

[4] Khalidi (2010).
[5] Laimann (2020).
[6] Mallon (2016).

that individual to satisfy those expectations. Because such relations are stable, Stability-Theorist succeeds.

Turn next to the upper-left corner of Table 7.1, where there is both reflexivity and fragility. Schizophrenia and autism are two of Hacking's own examples. Begin with schizophrenia.[7] At the start of the twentieth century, when schizophrenia was first named, its main symptom was flat affect. Auditory hallucinations, that is, hearing voices, were considered only a minor issue, not to be worried about, and not something to hide from the doctor. They were observed in many other psychiatric conditions, not just schizophrenia. The result was that hallucinations were increasingly reported by patients, and by the time a formal list of 12 symptoms of schizophrenia was compiled by Kurt Schneider 30 years later, the kind had changed, with hallucinations now designated the main symptom. But then, after the war, schizophrenia evolved again, from something viewed indifferently, even favorably, to a diagnosis that people wanted to avoid. Patients became less willing to report hallucinations. The definition of the kind changed, as hallucinations were gradually de-emphasized again as a diagnostic criterion (although they are still listed as one of the main symptoms). Schizophrenia is, thus, according to Hacking, a shifting kind because of reflexivity effects.

How will the schizophrenia kind shift next? That is hard to predict because it is hard to predict the social and political trends that determine attitudes to mental illness and to symptoms of schizophrenia. After the war, for example, something caused auditory hallucinations to be perceived as more shameful, but whatever that thing was, it was not predicted. The relevant relation is fragile. If it were not fragile then, like with the breeding of domestic dogs, Stability-Theorist could give us the answers we want. But no one has found a master theory that explains the past or will predict the future of the schizophrenia kind reliably—no equivalent to Darwinian theory. In-depth local investigation is required instead, such as the history that Hacking recounts. Fragility explains why.

Similar remarks apply to autism.[8] First named in 1938, this kind has subsequently varied greatly in its definition, as well as in theories of what causes it and in its degree of stigma. Reflexivity effects are an important part of the story, according to Hacking. The fragility of the relations that have determined the history of autism is revealed by the need for detailed local investigation to explain this history and to predict autism's future.

---

[7] Hacking (1999, 112–4).
[8] Hacking (1995).

Turn now to the two right-hand boxes. As cases of nonreflexive kinds, they would not usually feature in discussions of reflexivity. But optimal methodology differs between them, in the same way as it does between the two left-hand boxes. This difference is revealing because fragility, not reflexivity, is what tracks it.

In the upper-right corner, there is fragility but not reflexivity. Start with species invasions. The relevant kinds here are things like tree species, soil nutrients, islands, and lakes, and in the context of species invasions, none of these kinds shifts or is reflexive. But although we know several mechanisms behind species invasions, we cannot predict reliably which of them will apply (see Chapter 5).[9] These mechanisms are fragile. Stability-Theorist does not work: attempts to build a grand model to cover all species invasions have proved unproductive. Progress is made instead by contextual investigations, as per Case-Worker.

The World War One truces are another example of fragility without reflexivity.[10] Stability-Theorist, in the form of the Prisoner's Dilemma game, is not successful (see Chapter 3). Again, Case-Worker is required instead, as exemplified by the investigations of the historian Ashworth. The relevant relations are fragile. For example, when British and German soldiers were stationed opposite each other for a prolonged period, would that lead to spontaneous truces? Sometimes yes, sometimes no. To know which, each time further investigation is needed.

In the lower-right corner, finally, there is neither fragility nor reflexivity. Without fragility, Stability-Theorist succeeds: capturing the cogs and wheels of nature is again the efficient route to prediction, intervention, and explanation. A paradigm case is Newtonian theory. Another is artifacts. Electric toothbrushes, for example, are engineered to exploit relations—such as between pressing a button and a motor turning on—that, in the shielded environment of the toothbrush, are stable. Newtonian theory's kinds are typically nonreflexive, as are those of artifacts. But when it comes to choosing between Stability-Theorist and Case-Worker, it does not matter whether kinds are reflexive. It matters only whether relations are fragile.

---

[9] Elliott-Graves (2016, 2018, 2019).
[10] Northcott and Alexandrova (2015).

7.2 Laimann and beyond

Jessica Laimann's recent, penetrating discussion of reflexivity shares many of the above emphases.[11] How does a fragility perspective complement and add to it?

According to Laimann, our concern with reflexivity is ultimately epistemic and methodological. She writes (italics added): "Only when we *understand the mechanisms* that support patterns of change and stability among the members of a kind are we in a position to *provide accurate explanations and make inductive inferences* across a variety of contexts."[12] Shifting kinds are not necessarily a problem. What matters is the mechanisms behind the shifts: how well do we understand those? We can "provide accurate explanations and make inductive inferences" only if we can predict when those mechanisms hold—in other words, only if they are *not* fragile. Laimann also writes (italics added): "The problem with human interactive kinds is not merely that the classified objects change, but that they change in ways *unforeseen by our extant theoretical understanding of the world*."[13] Fragile relations, by definition, lead to changes that are unpredictable without supplementary knowledge, in other words precisely to changes that are "unforeseen by our extant theoretical understanding of the world."

As Laimann points out, the question of how quickly kinds change is a red herring. For example, many bacteria change their nature rapidly, but they may still be analyzed successfully by Darwinian theory. Gender kinds, in contrast, do not change at all because of stabilizing social effects, according to Mallon, but we nevertheless predict, intervene, and explain wrongly if the underlying social relations are not understood. What matters is not how quickly kinds change, but whether relations are fragile.

A central claim of Laimann's paper is that human kinds are often hybrid in a particular way. They have a dual nature: the properties that explicitly define the category (the base kind), but also the social position an individual occupies or the social role the individual plays in virtue of being recognized as a member of that category (the status kind). Laimann gives the example of sex as a biological base kind versus gender as a social status kind. In much everyday and scientific speech, "man" and "woman" are hybrid kinds that encompass

---

[11] Laimann (2020).
[12] Laimann (2020, 1056).
[13] Laimann (2020, 1051).

both aspects. Often, the base kind is nonchangeable while the status kind is changeable (although perhaps not in the case of gender, as mentioned earlier).

The fact that many human kinds are hybrid leads to two difficulties, Laimann argues. The first difficulty she calls *biased conceptualization*. This is when the status element in a kind is ignored, with the result that, surprised by it, our predictions and explanations go wrong. For example, if schizophrenia is treated purely as a symptom profile or purely as a neurological condition, then we would miss, according to Hacking, how people diagnosed as schizophrenic are singled out for certain expectations, opportunities, and treatments, and how this in turn leads to a change in the behavior of schizophrenics and thus, eventually, to a change in the definition of the kind itself. As Laimann writes, if we conceive of a hybrid kind "solely in terms of the base kind, without considering the associated status, causal pathways associated with the status disappear out of sight."[14] The result is fragility. If we know schizophrenia only by its current definition in terms of symptoms or neurological features, we are left unable to predict future changes in the kind reliably.

The second difficulty, according to Laimann, is *not understanding social status effects*. Social mechanisms are many and complex, and their effects, or whether they are even operating at all, are often difficult to predict. In other words, social status effects are often the products of fragile relations. Unlike biased conceptualization, not understanding social status effects is not a conceptual error, because even when we recognize the true nature of hybrid kinds, still the social science of them can be difficult. Laimann gives the example of the rise of the gay rights movement in the United States after the Stonewall riots in 1969, which greatly and rapidly changed the status kind component of "homosexual." This event was the result of a unique constellation of social and political events. It was hard to predict and remains hard to fully explain.

A key point for Laimann is that reflexivity is not the key feature. Both biased conceptualization and failure to understand social status effects are at root a deficit in our knowledge of the causal relations behind the social status aspect of hybrid kinds. If only we had this knowledge, then we could predict and explain successfully, regardless of reflexivity. I agree with Laimann. A fragility perspective allows us to pinpoint exactly what this deficit in our causal knowledge is.

Laimann convincingly shows one route—hybrid kinds—by which social sciences fall prey to fragility. We may add to this that there exist other routes to fragility besides hybrid kinds and that fragility is not unique to social science.

---

[14] Laimann (2020, 1060).

7.3 Reflexivity, good and bad

Does a focus on reflexivity help or hinder? On the positive side, reflexivity can be a useful indicator of fragility. Laimann's mechanisms of biased conceptualization and of not understanding social status effects are two ways this can happen. As an indicator, though, reflexivity is fallible. As we saw, sometimes reflexivity comes without fragility; other times, fragility comes without reflexivity.

Awareness of reflexivity also brings a second benefit. It can alert us to mechanisms that *reduce* fragility. A familiar case illustrates: the self-fulfilling prophecies behind bank runs. Banks' cash reserves typically cover only a fraction of their depositors' credit, so if all depositors demanded their money simultaneously, then the bank would face a liquidity crisis. In normal times, this does not happen. But rumors or reports that a bank is in trouble can spur all depositors, made worried about the bank's solvency, to withdraw their money at the same time. Thus, mere rumors of trouble can cause actual trouble—even if, initially, the rumors are false. This is reflexivity in action: the analysis of a target, in this case the rumors about the bank's solvency, itself influences that target. Knowing the mechanism of the self-fulfilling prophecy allows us both to predict and to explain the bank run, as shown by many historical examples. But it does more: it guides us toward interventions that *prevent* bank runs, such as granting regulators the power to prevent deposit withdrawals or to guarantee all deposits up to a certain value. These measures are designed to allay depositors' fears of a bank run losing them their money, and thereby to prevent the run in the first place. Preventing bank runs in this way means that relations such as that between depositing money in a bank and having access to that money later are rendered reliable. That is, knowledge of reflexivity enables stability.

A similar story is true of many other rational expectations economic models. Awareness of reflexivity enables us to turn some relations from fragile to stable.

Is reflexivity an effect of fragility or a cause of it? It can be either. Some fragile relations are causally upstream of reflexivity, others are causally downstream. For example, in a bank run, will authorities intervene effectively? Suppose that this is hard to predict. Then the following relation is fragile: a bank being rumored to be in trouble causes the authorities to intervene effectively. In turn, because of this fragility, depositors lack reassurance, and rumors of trouble become self-fulfilling prophecies. That is, here fragility causes reflexivity. But matters do not stop there. For the occurrence of bank runs then causes a new relation to become fragile, as noted above, namely the relation between depositing money and having access to that money later. Having been caused by one case of fragility, reflexivity then causes a new case of fragility.

These are the positive sides for science of a focus on reflexivity: it can be an indicator of fragility, either as a cause or effect of it, and it can illuminate some mechanisms and thus license useful interventions. Turn now to the negative side.

The main danger is simply misdirection: we should focus on fragility because fragility matters more. Reflexivity is a distraction. It is merely one source of fragility, and not an especially important one. An emphasis on how to define human kinds is also a distraction, at least with regard to methodology, as human kinds are far from the only kinds that partake in fragile relations.

But there is, in addition, another danger: a mistaken skepticism about social prediction. For in its more radical forms, an emphasis on reflexivity denies that systematic predictive success in social sciences is possible at all. This skepticism is a priori. Given free will, the argument runs, humans are always free to react to any prediction about themselves in such a way as to falsify it. Suppose, for example, that an unpopular candidate is predicted to win an election because of low voter turnout for their opponent. This prediction might itself inspire the previously apathetic supporters of the opponent to vote and thereby prevent the unpopular candidate from winning—in which case, the prediction falsifies itself. Because prediction about humans is always vulnerable to reflexivity effects, the argument runs, it is inevitably unreliable. Therefore, the argument concludes, in social science, we cannot use prediction to test scientific theories in the same way as we can in natural science. We cannot use it as a basis for action, either.

This skepticism has had distinguished proponents. They include Hayek, Popper, and MacIntyre; many interpretivists; and intellectuals outside academia, such as George Soros, Michael Frayn, and Jonathan Miller.

Some responses to this skepticism, such as John Stuart Mill's, claim that free will is compatible with determinism. But whether it is or not, a better response, I think, is to point out the obvious fact that social predictions are often successful. The interesting issue is when and why they are.

No doubt, social prediction is challenging. Prediction generally is challenging. But from a methodological point of view, reflexivity is merely one source of fragility, and it does not make third-person causal investigation impossible any more than other sources do. As Laimann emphasizes, a moving target is not itself a problem if we understand what determines the moves. If we do, then we may still predict perfectly well, as we do with domestic dogs and gender roles. Reflexivity does not somehow magically negate this. A priori arguments that it does are falsified by ample experience.

Skepticism about social prediction is not just misguided. It is also pernicious. Why? Because, in effect, it denies social science the possibility of empirical testing, which is the key to advancing knowledge. Here is one example. In a UK government press conference in April 2020, Health Secretary Matt Hancock was asked whether total UK COVID-19 deaths could still be kept below 20,000—the official figure had just reached 10,000. He replied (italics added): "The future path of this pandemic in this country is determined by how people act. That's why it's so important that people follow the social-distancing guidelines. *Predictions are not possible, precisely because they depend on the behaviour of the British people.*"[15] Here, Hancock explicitly endorses skepticism about social prediction. At one level, his statement is simply one of epistemic humility, correctly noting that part of the causal chain that would determine case numbers was the public's behavior. But the statement, if accepted, would also make it impossible to hold policy to account: it implies that we cannot fairly assess whether a prediction of a particular policy's effect is rational and, thus, whether the policy is worthy of praise or blame. No prediction can be deemed better than any other. Responsibility for the outcome is conveniently evaded—and put on the public instead.

Of course, Hancock understandably wanted to maximize public following of restrictions, and so had reason to emphasize the importance of that rather than of a prediction. Perhaps telling the public that the outcome depended on its own behavior was merely in the service of this urgent practical imperative, and Hancock himself did not believe that social predictions cannot be fairly evaluated. But what Hancock himself believed is beside the point. The problem is that skepticism about social prediction gives cover to such evasions of responsibility, and this is pernicious, both epistemically and morally.

Denying social prediction is also philistine: it denies that much actual, successful scientific inquiry is even possible. From where does this philistinism come? Here is one suggestion. The root cause of the error is a mistaken focus on reflexivity rather than fragility, which in turn arises from an agenda that is ultimately external—drawn primarily not from philosophy of science or from science itself, but instead from wider philosophy. This can be seen in many traditional handbooks, anthologies, and introductions to philosophy of social science. Typical questions are "what is intentional explanation?" and "how may we causally explain human action?," which reflect the agendas of philosophy

[15] BBC (2020).

of action and philosophy of mind; "is there collective agency?," which reflects the agendas of metaphysics, ethics, and philosophy of action; and "do special sciences reduce to physics?," which reflects the agendas of metaphysics and philosophy of mind. The same questions drive much of the attention given to reflexivity. But none of them is primarily motivated by what methods make social science successful or unsuccessful, where this is measured in the currency of predictions, (third-person) explanations, and interventions. For that, we should focus instead on fragility.

7.4 Natural versus social science

Reflexivity is often cited as a key respect in which social and natural science are fundamentally different. In fact, it is not clear that reflexivity is unique to social science—in addition to domestic dogs, arguably there are other cases from biology as well.[16] But even if reflexivity were unique to social science, reflexivity would not imply any significant divide methodologically, because what matters methodologically is not reflexivity but fragility.

What of natural versus social science generally? It is the dichotomy between fragility and stability that explains why features claimed to be typical of social science have the methodological implications they do. What are these features?[17] Social science is obliged to study concepts we care about, rather than choose only kinds that are projectable or that behave nicely; many of these concepts are difficult to define precisely and manifest differently in different settings; and many are socially constructed, so their behavior is governed in part by convention, legislation, and habit. All of this makes it more likely that relations involving these concepts are fragile. Further, social science is usually expected to deal with complex, open systems that are not easily shielded. In Nobel laureate economist Trygve Haavelmo's words: "Physics has it easy. No one asks physics to predict the course of an avalanche. But economists are expected to predict the course of the economy."[18] Again, this makes it more likely that relations of interest are fragile.

For these reasons, there might well be a correlation: relations of interest in natural science are more often stable and those in social science are more often fragile. That would explain why some stability-friendly methods, such as the

---

[16] Cooper (2004).
[17] Cartwright et al. (2022, 195–6) gives one list, from which I draw here.
[18] Quoted in Cartwright et al. (2022, 196).

search for laws or the use of laboratory experiments, are less successful in social science. One recent survey shows that social scientists are less realist about theories than natural scientists are, which is consistent with relations in social science being fragile more often.[19] And as discussed in Chapter 5, some a priori arguments claim that social science is more prone to computational irreducibility, which would again make fragility more common.

But the correlation is not perfect. Many relations in natural science are fragile, such as the mechanisms behind species invasions. And there are counterexamples the other way, too: relations in social science are sometimes stable. One famous example is the law of demand (see Chapter 5). Another is Durkheim's analysis of the causes of suicide rates—not coincidentally, discovering this stability also led Durkheim to pioneer the use of statistics.[20]

As with reflexivity, much of the philosophical interest in the natural–social science dichotomy has not been rooted in philosophy of science. Instead, it reflects the agendas of metaphysics or philosophy of action: according to some in those fields, a fundamental cleavage exists between the human and nonhuman worlds. But with respect to fragility, there is no such cleavage.

---

[19] Beebe and Dellsén (2020).
[20] Durkheim (1951/1897).

8

Fragility and economics

8.1 Theory monism and the efficiency question

In this chapter, I examine how fragility bears on economics and philosophy of economics. Economic theory, used well, has immense value. Yet it has been much criticized; fragility illuminates which criticisms hit home.

Before turning to economics, begin with some wider groundwork. We have seen that, when relations are fragile, theories play a toolbox role. We have also seen that theories therefore lose exclusivity. Realist interpretation of a theory—as opposed to specific applications of it—is no longer appropriate, and there is no prospect of a unique true theory to be worked with exclusively. Theory monism loses support. Instead, we should be liberal: many different approaches, formal and informal alike, can play a toolbox role usefully.

Fragility therefore raises an *efficiency question*: how should resources be allocated between different approaches?[1] One example of this question arises in Chapter 3, where I criticized the Prisoner's Dilemma literature for its inefficiency: it does not deliver enough explanations and predictions in return for the resources it has soaked up. The underlying problem is a Stability-Theorist strategy of theory development that is inappropriate to fragility.

Wasted scientific effort is easier to identify when relations are stable. If a theory is working well, we can intervene effectively, explanations and accurate predictions accumulate across a wide range of cases, and developments and elaborations of the theory are vindicated empirically. And if a theory is not working well, the need for a change of course becomes apparent quickly and clearly. Things are more difficult to judge, though, when relations are fragile. Progress achieved through Case-Worker is not built on the back of one theory. And because progress is now more piecemeal and patchwork, it takes longer to identify and weed out unproductive theories and research programs. The upshot is that, with fragility, scientific effort is at greater risk of being wasted, and the efficiency question becomes more urgent.

---

[1] Northcott (2018).

Science for a Fragile World. Robert Northcott, Oxford University Press. © Robert Northcott 2025.
DOI: 10.1093/9780191944352.003.0008

8.2 Example of theory monism and the efficiency question: economics

With the framework of the efficiency question in place, turn to this chapter's main subject: orthodox economic theory.[2] First, a note on terminology. "Economic theory" is a general orientation or research program as much as a single theory. I will understand it to mean the style of the neoclassical models that dominate mainstream economics, that is, mathematical models that deduce the equilibrium outcomes of interactions between rational agents. "Rational" here means utility-maximizing, although these days insights from behavioral economics are often added when empirically helpful.[3] True, it is specific models, not general theory, that make explanatory claims about real-world targets. But these models are constructed in accordance with the dictates of the wider research program, so we may speak of the predictive and explanatory record of the latter.

I will analyze economics as a case of theory monism. Orthodoxy is dominant: heterodox approaches are found rarely, for example, in the discipline's most prestigious journals. True, the orthodoxy was more dominant a couple of decades ago. Since then, behavioral economics, experimental economics, "modern monetary theory," and, most notably, the so-called empirical turn (discussed below) have all risen in prominence. Nevertheless, economics remains much more monistic than other social sciences.

Many economists have taken to heart the ambition of a single framework to underpin, and hence to unify, not just economics but all of social science. The prospect of a theoretical monoculture is celebrated. Indeed, it is seen as essential—the alternative is condemned as fragmentation that cannot add up to anything serious. In this, they follow the fierce spirit of Imre Lakatos, who (speaking about social psychology in his day) opposed:

... patched-up, unimaginative series of pedestrian "empirical" adjustments ... Such adjustments may ... make some "novel" predictions and may even conjure up some irrelevant grains of truth in them. But this theorizing has no unifying idea, no heuristic power, no continuity. They do not add up to a genuine research programme and are, on the whole, worthless.[4]

[2] Parts of this section are adapted from Northcott (2018).
[3] Angner (2019).
[4] Lakatos (1970).

Only one approach is deemed admissible. Economist Edward Lazear:

> ... economics stresses three factors that distinguish it from other social sciences. Economists use the construct of rational individuals who engage in maximising behaviour. Economic models adhere strictly to the importance of equilibrium as part of any theory. Finally, a focus on efficiency leads economists to ask questions that other social sciences ignore. These ingredients have allowed economics to invade intellectual territory that was previously deemed outside the discipline's realm.[5]

Nobel laureate Robert Lucas:

> Like so many others in my cohort, I internalized [the] view that if I couldn't formulate a problem in economic theory mathematically, I didn't know what I was doing. I came to the position that mathematical analysis is not one of the many ways of doing economic theory: it is the only way. Economic theory is just mathematical analysis. Everything else is just pictures and talk.[6]

And Jack Hirshleifer, from a special issue of the discipline's flagship journal: "there is only one social science ... economics really does constitute the universal grammar of social science."[7]

The confidence and ambition behind these and similar quotations have been reflected in a burgeoning program to expand economic-style analysis to topics previously thought noneconomic, such as racial discrimination, crime, and marriage.[8] Nothing is out of bounds to a framework that is universal.

One symptom of this theory monism has been the neglect of methods associated with nonorthodoxy. Examples include interviews and ethnographic observation; questionnaires; focus groups; discourse analysis; conversation analysis; small-N causal inference, such as qualitative comparative analysis; purely predictive models; causal process tracing; machine learning from big data; historical studies; randomized controlled trials; laboratory experiments; and natural and quasi-experiments. As noted, some of these latter methods have begun to be co-opted by the mainstream, but they remain much more widespread in other field sciences.

---

[5] Lazear (2000, 99).
[6] Lucas (2001, 9).
[7] Hirshleifer (1985, 53).
[8] Becker (1976).

8.3 Against theory monism in economics

Has all this been harmful? Should economists do orthodox modeling, or should they invest their efforts elsewhere? This is an efficiency question. To answer it requires, so to speak, an epistemic cost–benefit analysis. The costs are the resources invested into orthodox modeling, such as mathematical training of students, and perhaps more notably the opportunity costs, such as fieldwork methods not taught and fieldwork not done. The benefits are the successful explanations, predictions, and interventions that orthodox modeling leads to. What is the right balance? The question is not whether orthodox models *can* be useful—clearly, they can—but rather how *efficiently* they are useful.

Of course, such an epistemic cost–benefit analysis can be done only imperfectly and approximately. It is hard to count explanations and predictions in an objective way, hard to weigh these versus other goals of science, and hard also to evaluate the counterfactual of whether a different allocation of resources would do better. But implicitly, efficiency analyses are unavoidable and are being done already—every time a researcher chooses, or a graduate school teaches, one method rather than another, or journals or prizes or hirers choose one paper or candidate rather than another.

A presupposition of such a cost–benefit analysis is that alternatives exist. One demonstration of that: other social sciences show a different way. Della Porta's analysis of violent political groups, for example, prospered precisely because it rejected a micro-foundations rational-choice approach (see below) and instead developed a middle-level, less formal theory.

As noted, another example of an alternative comes from within economics itself: the recent, widely remarked "empirical turn." Between 1983 and 2011, in the five most prestigious economics journals, the percentage of papers that are purely theoretical—that is, free of any empirical data—fell from 57% to 19%.[9] Empirical work in prestigious venues is now more often a-theoretical, in the sense that it establishes previously untheorized causal relations rather than tests already-existing models.[10] Citation numbers suggest that the a-theoretical work is at least as influential. And in a systematic survey of over 130,000 articles from 80 economics journals, the proportion of papers that are empirical increased between 1980 and 2015, as did the proportion of citations that were to empirical papers. These increases were even greater if journals were weighted

[9] Hamermesh (2013).
[10] Biddle and Hamermesh (2016).

by prestige.[11] And anecdotal evidence tells a similar story, such as the research methods of recent winners of the John Bates Clark Medal.

An obvious possibility is that the empirical turn has been caused by, in effect, accumulated cost–benefit analyses by practitioners. These have motivated a shift from theory to empirical application.

The empirical turn's mere existence shows that theoretical and empirical work are distinct enough that we may speak of a shift in resources from one to the other. It also reaffirms that a discipline's norms and incentives are not too entrenched to preclude such a shift—the status quo is not inevitable. The variety of practices across different sciences shows the same thing.

Many others have advocated against economic orthodoxy. The objection in this book, to repeat, is that if relations are fragile, then the realist motivation for theory monism is defunct, and in economics, relations often—even typically—*are* fragile. We face an efficiency question: is theory monism a good allocation of resources?

One symptom of theory monism has been the demand that all macroeconomic models have *micro-foundations*, that is, that they must be based on formally modeled interactions between economically rational agents. Initially, the main target of this demand was Keynesian theory, which postulates relations between macro-level aggregates such as output, demand, and money supply, without any account of how these relations result from underlying interactions between individual economic agents. Such models therefore breach the tenets of monist orthodoxy. Besides a crude methodological reductionism, it is hard to see what could motivate insisting on micro-foundations—other than commitment to theory monism. So, if we reject monism, we should reject an insistence on micro-foundations.

(Others reject an insistence on micro-foundations for other reasons—because there are emergent higher-level phenomena in economics,[12] because of the psychological evidence that contradicts orthodox micro-foundations at the individual-agent level,[13] or because micro-foundations are not worth the extra complexity.[14] These critiques are independent of mine here.)

Many have criticized economic orthodoxy's handling of social-level influences such as culture or institutional structure or of situations in which human behavior predictably contradicts economic rationality. It has even been alleged that economic orthodoxy is inevitably hopeless, on the grounds

[11] Angrist et al. (2017).
[12] Epstein (2014, 2015); Hoover (2009).
[13] Hausman (1997).
[14] Ruiz and Schulz (2023).

of its idealization, or its ontology, or because of rational choice theory's dubious foundations.[15] But these criticisms miss the orthodoxy's successes. The real worry is one of balance. The foil is not economic orthodoxy, but its dominance.[16]

Many philosophers of science support such calls for pluralism. Hasok Chang writes: "if we should find a field of science that is quite monistic, then that is quite likely not healthy, and we should consider reforming it."[17] Nancy Cartwright and coauthors write: "a good precisely-specified method can establish only results of a certain *kind*. So, if you privilege some set of methods ... you necessarily thereby circumscribe the kinds of results that can be established."[18] A mixed diet is better. In Adrian Currie's phrase, we should be "methodological omnivores."[19]

But it might be objected that many practitioners in economics already advocate a version of Case-Worker, and thus of pluralism. Dani Rodrik, for example, influentially advocates that we accumulate a toolbox of models and then select from this toolbox case by case.[20] He explicitly opposes any search for a single universal model. And Nobel laureate Esther Duflo has advocated the "economist as plumber," who seeks to attain causal knowledge case by case, again in opposition to a single universal model.[21] So, are we attacking a straw man? Has Case-Worker taken over already?

Alas, I think not. Rodrik does indeed support, in effect, Case-Worker for theory *application*. According to him, no model applies universally. Instead, "economics advances by expanding its library of models," and the empirical details of the case should determine which model we apply.[22] All this is to the good. But Rodrik has a different attitude to theory *development*. He insists that only orthodox models belong in the library—while pluralist about models, he is monist about wider theoretical approach. He is explicit that economic models should be mathematical. And he is explicit that unorthodox modeling should be disregarded. He notes approvingly: "because economists share a language and method, they are prone to disregard, or deprecate, non-economists' points of view. Critics are not taken seriously ... unless they're willing to follow the rules of engagement. Only card-carrying members of the profession are

[15] Elster (1988); Rosenberg (1992); Nelson (1990); Lawson (1997).
[16] Northcott (2022c).
[17] Chang (2012).
[18] Cartwright et al. (2022, 33).
[19] Currie (2019a).
[20] Rodrik (2015).
[21] Duflo (2017).
[22] Rodrik (2015, 5).

viewed as legitimate participants in economic debates."[23] I diagnose this narrowness to be a residue of Stability-Theorist. That is what motivates using only a limited range of building blocks to develop our theoretical corpus, in abstraction from continuous empirical refinement. What else could justify restricting ourselves like this?

It is a similar story with Duflo. Many aspects of her methodology are admirable. But she is not pluralist about methods—she greatly favors randomized field trials for causal inference—nor, like Rodrik, is she pluralist about the range of hypotheses to be tested in those trials, which she takes overwhelmingly from economics. It is a truism that any empirical work presupposes *some* "theory" in the form of background assumptions. The issue here is whether these background assumptions must include those of economic orthodoxy.

Arguably, the most useful economic models are also the simplest: examples include the law of demand or basic considerations of opportunity cost or of incentive compatibility. This is no accident. As discussed in Chapter 4, when relations are fragile, Stability-Theorist theory development risks rapid degeneration into a house of cards. In a slogan, we should expect *diminishing returns to sophistication.*

Lack of pluralism in economics has a long history. Lionel Robbins influentially defined the discipline back in the 1930s as "the science which studies human behaviour as a relationship between ends and scarce means which have alternative uses."[24] This definition has proved prescient. In tandem with the rise of orthodoxy, economics—unlike most sciences—has come to be defined by its style of theorizing, rather than by the subject matter this theorizing is applied to, such as money, jobs, and factories. But economic relations are often fragile. This favors defining economics in terms of subject matter, contrary to Robbins, because that is the better way to pick up the contextual knowledge central to Case-Worker. And it tells against defining economics in terms of a single style of theorizing because when relations are fragile, such monism should be avoided. The continued popularity and relevance of Robbins's definition is a bad sign because it is a symptom of monism.

Perhaps it is wrong to interpret Robbins's definition of economics as picking out a method rather than subject matter. Instead of money, prices, and recessions, this thought runs, Robbins simply switches our focus to a different subject matter: constrained choice behavior, which, it turns out, we find in many places outside traditionally economic domains. But in practice, this

---

[23] Rodrik (2015, 80).
[24] Robbins (1935, 16).

new "subject matter" has meant that only a narrow style of theorizing is employed, and only one aspect—namely, constrained choice behavior—of target situations is modeled. This mistakenly leads us away from historian-like, contextual investigation of other aspects. So, whether we interpret Robbins as specifying a style of theorizing or as specifying merely a subject matter, either way, we are pushed away from Case-Worker.

It has been argued that the economics Nobel Prize favors new methods or techniques rather than discoveries of real-world facts or explanations.[25] If so, this represents a focus on style of theorizing rather than subject matter, the same mistake as in Robbins's definition. The natural science Nobel Prizes are not like that.

There is one other, classic defense of orthodoxy in economics: Milton Friedman's claim that the truth of assumptions does not matter.[26] Real humans are not always economically rational, Friedman concedes, but what is important is not an economic model's assumptions but only whether it predicts well. This defense is a form of instrumentalism—with respect to predictions of target phenomena (rather than to all the behaviors of individual agents).

The appropriate response is "horses for courses." On some occasions, for some purposes, Friedman is right: it does not matter that assumptions are false if salient predictions and interventions are accurate.[27] This is supported by much work on idealized models in science in general. But at other times, for other purposes, as many have pointed out, Friedman's defense of orthodoxy runs into trouble. What if a model's predictions fail—as they often do? To adjust the model, we need guidance, and Friedman's black-box treatment of assumptions gives none. In Daniel Hausman's phrase, it pays to "look under the hood."[28] The same applies to extrapolation: in which new context might a model work well, and in which not? Again, going black-box cannot help. Just as urged by the mechanistic view of explanation, it is fruitful to understand *why* a relation holds, not just *that* it holds.

Friedman's position is a defensive move. I think it is unfortunate—but not because economic orthodoxy's idealizations are always baleful. The real problem is theory monism. One consequence of this monism is economic orthodoxy's notoriously spotty predictive record, even with respect to target variables—thereby failing even by Friedman's own criterion. Another consequence of monism is that unorthodox models are ruled out a priori even

[25] Smith (2022).
[26] Friedman (1953).
[27] Herfeld (2018). See also Reiss (2012).
[28] Hausman (2007).

when they predict well. Again, this is objectionable even by Friedman's own criterion. These are the true difficulties here. The root cause of them is that monism is ill-suited to a domain in which relations are frequently fragile, and the biggest weakness of Friedman's defense of orthodoxy is that it does not engage with this root cause. Instead, it gives license to ignore it.

Finally, I have not discussed here a large debate in philosophy of economics, namely whether and how economic models explain. Why not? Because monism and the efficiency question are more pressing—they bear more acutely on how economic work is organized.[29] The explanation debate does not speak to them. The case that the Prisoner's Dilemma has given a poor return on investment, for example, is clear on any plausible view of explanation.

8.4 Against grand social theories

Label by *grand social theory* a universal framework for social science. Because relations of interest are frequently, even usually, fragile not just in economics but right across the social world, the arguments above against theory monism are also arguments against grand social theories. Yet the pull of such theories seems irresistible: many have been proposed. A highly sophisticated example is the late Herbert Gintis's influential 2016 book, *Individuality and Entanglement*.[30] A close examination is illuminating—not just of Gintis's book but also of the prospects for grand social theories generally.

I will be critical. But I choose Gintis's book as an example not because it is notably bad, but for the opposite reason—because it is notably good: a crisply written, dazzling synthesis, enormously knowledgeable about both theory and empirical background. Roughly, Gintis gives an evolutionary account of the origins and development of human society, including political systems and economic structure. Although Gintis's core tool is rational choice theory, he abandons methodological individualism by embedding his analysis in evolutionary game theory, adds in elements of behavioral economics, and in places also uses nonanalytic approaches such as complexity theory. Gintis presents the result as a unified conceptual scheme for all social sciences—a grand social theory. He is explicitly motivated by unification itself: "a scientific discipline attains maturity when it has developed a core analytical theory . . . sociological theory since Parsons has become unacceptably fragmented."[31]

---

[29] Northcott (2018); Alexandrova et al. (2021).
[30] Gintis (2016). Gintis (2007) advocates for the same project at article length.
[31] Gintis (2016, 138–9).

Gintis's framework, like a Kuhnian paradigm, offers not just a theory but also, implicitly, criteria by which to evaluate theory. This is a key point. A central goal for him is that target phenomena "can be modeled," by which is meant they can be mathematically derived from, or result from a simulation produced by, the master theory. In this way, a large range of phenomena can be represented in a single, unified scheme, which thereby provides a common language for all of social science. Examples of such phenomena in the book include the evolution of altruism, norms, and morality, of the social contract, and of property. There are also accounts of social class and general economic equilibrium, as well as, in passing, of the definition of life, scientific genius, and the ideas of Freud.

As is well known, and as Gintis remarks, rational choice theory is formally compatible with almost any observed action. This makes it an attractive foundation if our evaluation criterion is how much "can be modeled." But this criterion is clearly unsatisfactory. We need a higher hurdle, and for any science, a central one is surely empirical confirmation. To be sure, Gintis is sensitive to empirical findings, and he endorses the need for theory to be responsive to them. But in practice, this is not what drives his theorizing. Accurate predictions or interventions are largely absent. And although many explanations—and understandings and insights—are claimed, when relations are fragile, without direct empirical confirmation such claims are dubious, as we have seen (Chapter 3).

Gintis does mention empirical inadequacies from time to time. For example, he regrets that his theory has no account of the rise of gender and racial equality movements in the West, and he concedes that "the Walrasian system is a very poor guide to micro-modeling real economic transactions."[32] But notice that "micro-modeling" here is in effect code for "modeling that is empirically accurate," which is *contrasted* with the kind of modeling that a grand social theory does. Tellingly, Gintis is generally more concerned by theoretical shortfalls than by empirical ones. He highlights as an anomaly, not any empirical shortcoming, but rather when theory cannot model how equilibrium is attained: "the general economic equilibrium model is a static construct that gives no suggestion as to how equilibrium might be attained. This is a critical limitation."[33] When push comes to shove, the primary evaluation criterion is not empirical accuracy but theoretical comprehensiveness.

---

[32] Gintis (2016, 117).
[33] Gintis (2016, 121).

To make the problem sharp: how should we decide between Gintis's grand social theory and other grand social theories? By which of them has an analytically coherent, unified theoretical system? But there are many such systems, some of them coming from a similar tradition as Gintis.[34] By the range of target phenomena that "can be modeled"? But all candidates offer a large such range, and all of them offer "explanations" in the same sense that Gintis's system offers them. One evaluation criterion that is not emphasized is empirical confirmation, yet when relations are fragile, that is required continuously. Without it, or at least the prospect of it, the other virtues do not matter.

Compare Gintis's grand social theory with rivals from other traditions. Why prefer Gintis's account to a Marxist one? Or to those of Foucault, Bourdieu, Luhmann, interpretivists, political philosophers, or various political activists? Or, going further back, to those of Nietzsche or Hegel? Given the weakness of Gintis's empirical constraints, it is hard to rule out rival candidates, both actual and potential. The only reason to favor Gintis's theory is a prior preference for his style of theorizing, rather than any appeal to the usual empirical criteria of predictions, interventions, and (warranted) explanations. But in that case, in practice, a piece of work is evaluated according to whether its explanations can be derived from, or fit into, the grand-theoretical system, not according to whether they are empirically endorsed. That is, what ultimately matters is coherence with a prior conceptual scheme. I do not see how we are any longer doing science, as opposed to ideology.

Defenses do exist. One is "jam tomorrow." A grand social theory, this defense runs, does the hard work now of establishing a theoretical foundation, which, by unifying the field, prepares the ground for greater empirical progress in the future. Only in this way can we head off Lakatos's specter of fragmentation. This defense is yet another version of "one more heave." But it founders on two issues. First, underdetermination: many different grand social theories may appeal to the same jam-tomorrow defense. Without empirical success to guide us, how do we choose between them? Second, if relations are fragile, Stability-Theorist theory development is a poor bet for jam today or tomorrow.

A second defense appeals to grand social theories' heuristic value. Perhaps they highlight possibilities or categories that would not be thought of otherwise, and this is useful even when the theory itself lacks empirical warrant. Grand social theories do contribute in this way, sometimes. But the question is whether they do so efficiently: on balance, do the indirect benefits that grand social theories occasionally bring justify the narrow path they want to force

[34] Binmore (1994, 1998); Skyrms (2014).

on social science? I am dubious. The underlying reason is again that Stability-Theorist theory development is a poor bet when relations are fragile.

The replies to these defenses are illustrated by our case studies. Committing to the Prisoner's Dilemma framework, for example, actively impedes understanding of World War One—and other—truces. To develop the Prisoner's Dilemma yet further in the name of theoretical unification will not lead to jam tomorrow; it is a blind alley. The same is true with species invasions. Attempts to develop a single, unified model bring us no nearer to predicting or explaining individual invasions, and in fact, take us further away from that. Progress is made only by more piecemeal investigations.

Grand social theories instantiate the Stability-Theorist strategy par excellence. To do better, we must follow Case-Worker. Another of our case studies shows the way: della Porta's investigation of political violence in postwar Italy and Germany. This investigation succeeded precisely because della Porta did *not* abstract away from local detail in the quest for a universal theory. Instead, she developed theory in close connection with empirical feedback all the way, and because of that, her theory yielded warranted explanations, plus concepts and mechanisms for use elsewhere. From-first-principles rational-choice modeling was one of the rival approaches to political violence that did not yield these dividends, precisely because it abstracts away from crucial empirical details. That is not the way to go.

9

Fragility and big data

9.1 Big data and case studies

Might the number-crunching power of data-intensive science sweep away our dependence on all theories, fragile and stable alike, and usher in a new world: priority for prediction over explanation or causal understanding? Might it herald the "death of theory" altogether?[1] Will big data make fragility yesterday's problem?[2]

"Big data" is a vague label. Some interpret it narrowly to refer only to a few machine learning techniques. I will interpret it more broadly, in the spirit of the American Association of Public Opinion Research's definition: "an imprecise description of a range of rich and complicated set of characteristics, practices, techniques, ethical issues, and outcomes all associated with data."[3] A broad interpretation gives big data every chance to prove its worth. My conclusion—that its prospects as a prediction tool are limited—is then stronger. Although a huge range of techniques falls under big data so understood, these techniques have enough in common that their prospects may be assessed as a group.

Big data has chalked up impressive successes, with likely more to come.[4] Notable early examples include the discovery of the Clustered Regularly Interspaced Short Palindromic Repeats (CRISPR) technology for genome editing in living eukaryotic cells[5]; prediction of which manhole covers will blow, and which rent-controlled apartments will suffer fires, in New York City[6]; Facebook and Google's experiments with page design and marketing; and chatbots such as ChatGPT. New techniques are being developed apace. Neural nets, for example, are behind rapid advances in natural-language translation.[7] Do these successes presage a revolution?

---

[1] Anderson (2008); Hey et al. (2009).
[2] This chapter is adapted with permission from Northcott (2020).
[3] Japec et al. (2015, 840).
[4] Japec et al. (2015); Foster et al. (2017).
[5] Lander (2016).
[6] Mayer-Schoenberger and Cukier (2013).
[7] Lewis-Kraus (2016).

Science for a Fragile World. Robert Northcott, Oxford University Press. © Robert Northcott 2025.
DOI: 10.1093/9780191944352.003.0009

The heart of this chapter is four case studies of prediction of field phenomena: political elections, weather, gross domestic product (GDP), and economic auctions. Why case studies? As we will see, there exist general analyses already of what conditions are necessary for big data predictive methods to succeed. But when are these necessary conditions satisfied? And when they are satisfied, are they sufficient? There is no substitute for local detail—which means case studies. In each one, I will draw on previous work by others.

General analyses implicitly promise to shed light on big data's prospects in pressing actual cases. In this chapter, I proceed the other way around, so to speak. I start with pressing actual cases and then, informed by this dive into particularity, generalize out again to get a better sense of big data's prospects more widely.

Our examples come with no presumption for or against big data methods. The first three of them—elections, weather, and GDP—are of independent interest because of their importance. The fourth—economic auctions—is of independent interest because it is an example from social science of successful field prediction based on results from experiments, which is an influential method.

The argument in this chapter is generalist in that, as the examples show, the same factors are positives or negatives for big data's prospects across cases. It is also generalist in that these factors are positives or negatives across different big data methods. As it will turn out, the most important negative factor is fragility. Fragility is not yesterday's problem, after all.

9.2 First example: political elections

There are several ways to predict the results of political elections. By far the most successful is opinion polling.[8] The main alternative is to predict based on "fundamentals" that recur from election to election, most commonly macroeconomic variables such as growth in GDP, employment, or real wages. It is uncontroversial that polls predict better, but the alleged compensating advantage of models based on fundamentals is that at least they can explain election results, whereas polling cannot. I think that, in fact, neither approach can explain election results.[9] But we are concerned here only with prediction.

---

[8] (Northcott 2015) gives details and references.
[9] Northcott (2015, 2017).

Opinion polls use the voting intentions of an interviewed sample as a proxy for those of a population. How might things go wrong? The most familiar way is sampling error: small samples can be unrepresentative flukes. But sampling error is not the most important source of inaccurate predictions. Almost 25% even of late polls of US presidential elections, for example, miss by more than their official 95% confidence interval, yet the expected miss rate given sampling error alone is only 5%.[10] A far bigger challenge is sample *balance*. Results will be biased if a sample is unrepresentative of the voting population with respect to, for example, age, gender, race, or income, since these factors correlate with voting preference. This is different from sampling error. If a sampling procedure over-selects for men, say, that cannot be alleviated just by making the sample bigger.

Pollsters must decide exactly which factors to balance their samples for. Should they balance, for example, for declared political affiliation or for degree of interest in politics? This has been the source of errors in recent US and UK polling. Further decisions are needed too: how hard and in what way to push initially undecided respondents for their opinions, how hard and in what way to pursue respondents who decline to participate, whether to sample face-to-face or by phone or online and (in the latter cases) whether to interview or let respondents fill out answers alone, how to assess how firmly held a respondent's preference is, and how to assess the likelihood that a respondent will actually vote. Exactly how pollsters tackle such issues significantly influences the accuracy of their predictions.[11]

Separately from such internal issues, *aggregating* polls further improves predictive accuracy. One obvious reason is that aggregation increases effective sample size and therefore reduces sampling error. But mere aggregation is no cure for incorrect sample balancing because the optimal balancing procedure might not be the industry average. To minimize the chance that an aggregation of polls is systematically skewed requires a sophisticated second layer of method, distinct from that required to conduct a single poll.

With this backdrop in place: what role for big data in opinion polling? Clearly, more data have helped. Polling predicts better today in part simply because of more and better data; there were no reliable political polls at all until after World War Two. Improved analytical methods have also helped, polling aggregation being one example.

---

[10] Silver (2014).
[11] Sturgis et al. (2016); Wells (2018).

That is the past. What of the future? How much could prediction be improved: what is big data's upper limit? In the case of elections, the answer to this key question is, alas, that predictive paradise will remain elusive. Two difficulties are each individually sufficient to ensure that: not enough data and fragility. Let us see why.

Political campaigns increasingly use big data methods. These have mainly taken the form of "microtargeting": extensive data are collected about individual voters' consumption patterns, media preferences, demographic characteristics, and so on, and algorithms then track how these factors correlate with political preference and likeliness to vote. Barack Obama's 2008 US presidential campaign, for example, tracked over 800 different voter variables as early as the Iowa caucuses in January. Campaigning material and tactics are tailored accordingly, at the level of individual voters, in order both to change voter preferences and to increase supporter turnout. Such microtargeting, which first became prominent in George W. Bush's 2004 campaign, is an example of a "theory-free" big data approach displacing a more traditional model-based one. Might it enable campaigns, or anyone, to predict election results better than opinion polls do?

To achieve that, we would need to find correlations between the effect variable, that is, actual votes, and the putative cause variables, such as consumption patterns, media preferences, and so on. But there is an epistemological roadblock. The limited sample of past elections is not enough to train predictive algorithms, yet no other voting data are available because the secret ballot means that individuals' votes are unknown. Practitioners acknowledge generally that machine learning requires a lot more data than are available in most cases.[12] Elections are an example of this.

Why not simply *ask* individual voters how they voted? Some might answer falsely but, the reasoning goes, if most do not, we can establish correlations well enough to predict accurately. The salient question, though, is whether we can predict *better* with big data methods than we already do with opinion polls, and that is dubious if big data methods must rely on asking voters for whom they voted—after all, pollsters do this already. True, pollsters usually ask a voter how they *will* vote rather than how they *have* voted, but it is not clear that this makes answers less reliable. Indeed, it might be the retrospective answers that are less reliable because they can be swayed by post hoc perceptions of an election result, perhaps via a desire to be seen to have voted for the winning side or simply to be seen to have voted at all.[13] For big data methods to gain an

---

[12] Foster et al. (2017, 172–3).
[13] Issenberg (2016, 193).

advantage, it is not enough that, say, media preferences in isolation correlate with voting preference. Rather, balancing samples for media preferences must add predictive value over and above pollsters' existing balancing for other factors.

The benefit from this kind of augmented polling of voters is inevitably limited because it does not address the biggest difficulties that opinion polling faces. It is one thing to know what a voter's political preference is, but quite another to know whether they will vote. For example, polls in the 2015 UK general election were unusually inaccurate. A subsequent inquiry revealed the main cause: pollsters assessed likeliness to vote by asking voters themselves, but errors in voters' self-assessment correlated with political preference, which led to biased samples. Better to have used historical rates of turnout for demographic groups.[14] For the 2017 UK general election, therefore, most polling firms switched to the latter method. But their predictions were again unusually inaccurate. Investigation revealed, roughly, that the solution was to switch back to the 2015 methodology—the optimal method had reversed.[15]

This brings us, finally, to fragility. The UK polling failures illustrate a fundamental problem: the underlying causal processes were not stable. In this case, the causal relations between demographic variables and turnout changed between 2015 and 2017 in a way that could not be predicted. That is, they were fragile—or "nonstationary," in the parlance of practitioners. The problem is that this fragility cannot be overcome by knowledge of past correlations. The same is true of other electoral variables too: do, say, the percentages of Black people, women, the rich, sports fans, and so on, that vote for a particular party stay constant across elections? Historically, they have not.

One response to such fragility is to stick to short-term forecasting within a single campaign, in the hope that in this shorter timeframe, things remain stable. But even within a single campaign, there are many relevant fragilities, especially during primaries when voter preferences are more volatile. Problems arise during general elections too, as with temporary swings of opinion after notable events. And the effectiveness of a particular campaigning tactic can fade quickly with repeated use.[16] The fragility problem remains.

Two practices counter fragility. First, campaigns often run daily polls to help calibrate their inferences from data about consumer preferences and so on. This is a sensible tactic. But the accuracy of any election predictions inferred

[14] Sturgis et al. (2016).
[15] Wells (2018).
[16] Issenberg (2016).

from such daily polls still has an upper limit given by those polls' accuracy, so, again, there is no reason to expect outperformance of regular public polls.

A second way to counter fragility is for campaigns to utilize their extensive *non*polling data, namely, voters' responses to doorstep, phone, and other interactions. Such responses, in conjunction with polling, are important for calibrating a campaign's microtargeting algorithms and sometimes for altering them mid-campaign. These voter responses predict voting behavior, of course, but again what matters here is whether they predict better than regular polling does.

Similar remarks apply to the increasingly frequent use of randomized experiments and trials, which is another major innovation of recent campaigns.[17] Such experiments usually test campaigning tactics, with efficacy measured either by changes in turnout, which can be observed, or by changes in proxies for actual votes such as opinions expressed on the doorstep or in focus groups. Although these tactics can boost a campaign's number of votes, there is little reason to think they enable us to predict overall election outcomes better than polls do.

The latter point applies to the use of social media, too. Recent years have seen a vast expansion of campaigning of various sorts via social media, often in a data-driven way. But whatever the other impacts of this, it has not yet led to better prediction of election outcomes.

Perhaps it might be objected, the underlying fragility itself can be tackled by big data methods. This fragility is presumably a result of other causal processes, and these other processes might themselves generate trackable correlations. But this observation is not terribly helpful because it does not provide actionable advice beyond a truistic "look for variables that partake in important relations that are not fragile." And there is no guarantee that the remedy would work anyway. It is conceivable, even likely, that in hugely complex social domains such as elections, the data-generating processes are so fast changing that they never produce correlations that are both useful and trackable.

Returning to the real world, political campaigns certainly have ways to identify supporters and to estimate how likely they are to vote. But then, so too do opinion pollsters. Can campaigns predict overall election outcomes better than pollsters do? I am dubious. There is little more than fragmented, anecdotal evidence for that.[18] And in addition to the reasons above, there is evidence (admittedly some also anecdotal) against it: for example, all sides in the

---

[17] Issenberg (2016).
[18] Issenberg (2016, 324–5 and 348).

2016 US presidential election, the 2017 and 2015 UK general elections, and the 2016 UK Brexit referendum were privately surprised by the results. Campaigns themselves typically take opinion polls to be the best guides to who will win. And there is no indirect evidence of insider special knowledge, such as tell-tale activity on political betting markets.

Finally, there are two other alternatives, in some ways more in keeping with big data methods generally. First, should we just adopt an "n equals all" approach and ask *every* voter how they will vote? But such an interviewing marathon is not a realistic prescription. And even if it were, it would still be subject to familiar uncertainty over which respondents would vote and which were telling the truth.

Second, could we predict elections not by interviewing voters, but instead by tracking indirect indicators such as Google searches of candidates? Alas, this method's record is not encouraging, either for predicting elections or for predicting other phenomena such as flu outbreaks. Social media users are usually unrepresentative. And a further source of fragility is added, namely that Google's search algorithms themselves change frequently in ways opaque to outsiders.[19]

In sum, the prediction of elections has improved, and more data have helped. But only somewhat, because of the problems of not enough data (infrequent elections, secret ballots) and widespread fragility. Each of these problems alone is sufficient to thwart election prediction. They hamper all big data techniques alike, and that will likely continue.

9.3   Second example: Weather

Weather forecasting has improved greatly, as discussed in Chapter 6. Four-day forecasts today are as accurate as one-day forecasts 30 years ago.[20] Several factors together explain this achievement.[21] One is a huge increase in the quality and quantity of data since the launch of the first weather satellites in the 1960s: temperature, humidity, and other reports are of ever greater refinement, and tens of millions of observations per day are used in forecasts. Another factor is ever better forecasting models. A third factor is new analytical methods, notably since the late 1990s, the "ensemble method" of running

---

[19] Lazer et al. (2014).
[20] Mance and Sheppard (2023).
[21] Northcott (2017).

multiple simulations of stochastic models to generate probabilistic forecasts. Lastly, hugely increased computing power has enabled the other factors. The ensemble method of forecasting was previously infeasible because simulations could not be run quickly enough. The increase in data and computing power together enabled more sophisticated models to be developed. And additional data are not collected blindly; rather, experience of what kind of data most improves the accuracy of models' predictions has informed the choice of instruments on new satellites.

What role has big data played? Exploitation of more and better data is certainly one important ingredient, and methods such as ensemble forecasting count as data analysis techniques. Other big data techniques can be combined with theory as part of hybrid methods, which are useful for subsidiary tasks such as finding proxies for missing data or modeling clouds or vegetation.[22] On the other hand, improvements are not due to data factors alone. And they are limited. Even now, forecasts more than seven or eight days ahead cannot beat the baselines of long-run averages or of simply extrapolating current conditions.

Again, a key question is: how much *could* weather forecasting be improved in the future by big data methods? The availability of ever more data will help. Any new data must be collected by new physical instruments, which requires choices about which instruments to deploy and where, and such choices are in part informed by background theory. Still, for a given set of data, could weather be "blindly" predicted by machine learning techniques instead of by, as currently, a model adapted from physical theory? Attempts are underway, and they show some promise.[23] On the plus side, weather forecasting does satisfy the conditions that, we will see below, are necessary for machine learning techniques to succeed. Crucially, the relevant relations are stable: the Earth's weather system operates in roughly the same way from day to day and year to year, and we know and can measure those variables that partake in these relations. This is what enables the success of current forecasting models. On the minus side, if the weather system is chaotic, then given imperfect data, only probabilistic forecasts will ever be possible. For the data currently available, how accurate such forecasts could eventually become, how far in advance, is unknown.

---

[22] Knüsel et al. (2019).
[23] Mance and Sheppard (2023); Ravuri et al. (2021); Ham et al. (2019).

9.4 Third example: Gross Domestic Product

Predicting GDP has proved difficult.[24] One benchmark is to assume the growth rate of real GDP stays the same. Currently, 12-month forecasts out-perform this benchmark barely, and 18-month forecasts do not outperform it at all. Forecasts also persistently fail to predict turning points, that is, when GDP growth changes sign. In one study, in 60 cases of negative growth, the consensus forecast beforehand was for negative growth on only three of those occasions.[25]

The record shows little or no sustained difference in the success of different forecasters, despite widely varying methods. These methods include purely numerical extrapolations, both informal (chartists) and formal (usually uni-variate time series models improved by trial and error); nontheory-based economic correlations, both informal (indicators and surveys) and formal (multivariate time series); and theory-based econometric models, which sometimes feature hundreds or even thousands of equations.[26]

Unlike in the weather case, the forecasting record has not improved in 50 years despite vast increases in data and computing power. The induction is that more data and computing power will not make the difference. Given the complexity of what determines a country's GDP, likely no forecasting method captures all of the generating processes, and the generating processes that we can identify are fragile. These are crucial disanalogies with the weather case, and they bode ill for big data's prospects.

GDP forecasting faces several further difficulties that can lead to fragility (and that make prediction difficult even when they do not)[27]:

1) The economy is an *open* system, continuously impacted by noneconomic factors, such as election results, that do not appear in economic forecasting models. These make stable correlations less likely.

2) *Measurement errors* are large. GDP can be estimated only by aggregating meso-level inputs, which requires many statistical estimates and sub-jective judgments. Methods for seasonal adjustment introduce further imprecision. One symptom of these difficulties is significant discrepan-cies between different measurement methods. Another symptom is the

---

[24] See (Betz 2006) for details and references.
[25] Loungani (2001).
[26] Betz (2006, 30–8).
[27] Betz (2006, 101–8).

large size of revisions, which typically are greater than 1%—comparable to the average forecast error.

3)   The economy is a *reflexive* system. Forecasts can themselves affect the economy in such a way as to impede the task of forecasting it, which is why many rational expectations models deem it impossible to forecast systematically better than a random baseline.

4)   The economy might be a *chaotic* system, in which case at best only probabilistic forecasts are possible.

No big data method is a plausible solution to these difficulties.

9.5   Fourth example: Economic auctions

Laboratory experiments are increasingly common in social science.[28] They are also increasingly used to predict the impact of field interventions. Can big data methods help? I consider here one well-studied case, namely the US government spectrum auctions from the mid-1990s.[29]

This is a case study in Chapter 4, so I will resummarize it here only briefly. The task facing the auction designers was to sell many licenses across the country and to raise as much money as possible while also satisfying various noneconomic constraints. Game-theoretical models had revolutionized the auction literature in the 1980s. The intricate final spectrum auction design, though, was not derivable from game theory. An extensive program of laboratory experiments was required, carried out with human subjects, and mired in messy practical details, leading to many ad hoc adjustments and fine-tunings.

What prospects for big data methods? We have asked this question for each of our examples, but the auctions case reveals an implicit assumption behind it. Previously, the key to progress was mostly to collect existing data or to analyze that data better. But with the auctions, the relevant data had to be *created* by running experiments, and so in addition, it had to be decided what data to create. Prospects for prediction depended on those decisions.[30]

The type of predictive progress in the auctions case is different too. Much of the "data" relevant to predictive success was practical know-how, which by its nature tends to apply to new contexts only unreliably. This is why the spectrum

---

[28]   Kagel and Roth (2016).
[29]   Guala (2005); Alexandrova (2008); Alexandrova and Northcott (2009).
[30]   Leonelli (2016) emphasizes related work by "data curators" in systems biology.

auction in Switzerland in 2000 could fail even after the auctions in the United States several years previously had been a success: the relations that worked in the US case did not extrapolate reliably to the Swiss one; that is, they were fragile. Progress is piecemeal—predictive success in one task and then another task and then another, and so on. There is no trend of greater *degree* of predictive success, rather only greater *scope*.

The auction case suggests pessimism for big data's prospects. As is typical with fragility, success requires intricate contextual knowledge, in this case to create useful data. Progress was not a matter of better machine learning or data mining. If the auction example is indicative, big data methods will not transform the derivation of field predictions from laboratory experiments.

9.6 Conditions for big data predictive success

What determines when big data methods succeed? Both philosophers of science and practitioners have studied this. The surveys by Wolfgang Pietsch are a useful starting point.[31] He identifies four necessary conditions for success[32]:

1) Vocabulary is well chosen, that is, parameters are stable causal categories.
2) We have data for all potentially relevant parameters.
3) Background conditions are sufficiently stable.
4) There are enough instances to cover all potentially relevant configurations.

Label these the *Pietsch conditions*. They apply to big data methods generally, that is, to techniques of machine learning and data mining.

To see the need for Condition 1: suppose variable X perfectly correlates with Z, but Y does not. Then, we can predict Z by tracking X. But suppose instead we track only a composite variable X + Y. Then, we will fail to predict Z accurately, and so miss the chance to exploit X. In our four examples, though, satisfying this condition was not the relevant constraint.

The importance of Condition 2 is obvious: whenever full predictive success is absent, we cannot be sure this condition is satisfied. GDP and elections are clear examples. Condition 2 does not imply we need to know in advance which parameters are important—a requirement that would negate one of the great

---

[31] Pietsch (2015, 2016, 2021).
[32] Pietsch (2015, 910–11).

attractions of big data methods, which is that they themselves discover which parameters are predictively important. But they can do that only if they have the data.

Condition 3 refers to fragility. Any correlation is generated by some process. If that process varies unpredictably, then exploiting the correlation for prediction is thwarted. This problem severely limits the efficacy of big data methods for predicting GDP and, to some extent, elections. By contrast, the relative stability of the processes that generate the Earth's weather enables big data methods to be (potentially) more effective there.

Condition 4 is that the dataset must be sufficiently rich. Ideally, it should include all relevant configurations of cause-and-effect variables (although less than this ideal might still enable accurate prediction in a limited range). Satisfying Condition 4 is a problem in the elections case: the relatively small number of elections does not allow us to select between all of the many possible causal hypotheses. Again, weather is a contrast case, because we have ample records of every relevant combination of weather causes and outcomes.

In big data success stories, the Pietsch conditions are satisfied. The data-generating process is stable enough, the training set of combinations is rich enough, and the variables are well chosen enough, for us to predict accurately. For example, the process that causes some rather than other New York City manholes to blow seems to be stable, and enough data can be collected to identify the relevant correlations.[33] Thus, the Pietsch conditions do illuminate actual cases. Nevertheless, it is useful to augment them with case studies.

First, are the Pietsch conditions satisfied in important cases? They are in the weather example, but not in our other ones.

Next, why are the Pietsch conditions sometimes not satisfied? One reason is that a system is open, which threatens the stability condition because a generating process can be unpredictably influenced by external factors. If a system is open, this also threatens the condition that all relevant parameters are known (elections, GDP). A second reason concerns the sufficient data condition, which is threatened when too few iterations exist of the relevant event (elections), which in turn is sometimes caused by the relevant data being too contextual (auctions). A third reason is that a system is reflexive, as many social systems are, which can lead to fragility. It has been suggested that this applies to elections and GDP, although if it does, it is not clear how significantly.

As well as necessary for accurate prediction, are the Pietsch conditions also sufficient? One lesson of our examples is that they are not. The GDP case

---

[33] Mayer-Schoenberger and Cukier (2013).

illustrates further barriers to prediction, such as measurement error or if a system is chaotic. On the positive side, the examples also illustrate remedies for some of these difficulties. In the weather example, the ensemble method makes possible the probabilistic prediction even of a chaotic system. The weather and election examples also highlight that sometimes other factors too are necessary, such as the new techniques of ensemble forecasts and polling aggregation. Further, the distinctions between Pietsch's different conditions are themselves fuzzy, as sometimes the same issue may be categorized under more than one of them.[34]

True, these further difficulties may be recast simply as failures to satisfy Pietsch's conditions: an open system implies either fragility or incorrect vocabulary; reflexivity can lead to fragility; and measurement errors, unobservable variables, and too few events each implies that not enough accurate data are available. But then most of the real work would be assessing when and why the conditions apply, which means supplementary local investigation.

Prediction is often an amalgam of different methods, some of which rely on large datasets and use big data techniques, while others do not. So, big data's impact is not all-or-nothing. Even if it does not improve overall predictions, it can still help with local modeling tasks. It has been argued, for example, that machine learning techniques can detect whether data-generating processes are stable.[35] If so, these techniques can detect fragility even if they cannot cure it. In climate science, too, big data methods help even when they are not a panacea for prediction.[36] And the same is true in our examples. Political campaigns, for instance, successfully use big data methods to predict and influence many aspects of voter behavior, even though they cannot predict overall election results any better than pollsters do.

A recent advance in election forecasting is multilevel regression and poststratification (MRP) modeling. This breaks down polling data by demographics and then combines that with demographic information about individual districts to construct predictions at a district level that would otherwise require large polls in every district. The method is complicated to implement but significantly improves forecasting accuracy. It does so, first, by improving on a simplistic assumption of a uniform swing between political parties across a whole country, and second, by ameliorating the impact of incorrectly weighting a sample by demographic factors. Why is MRP modeling relevant?

[34] Northcott (2020, 102).
[35] Chen (2021).
[36] Knüsel et al. (2019).

First, because it instantiates the Pietsch stability condition, MRP modeling works only if the relation between demographics and voting behavior is constant across districts. And second, because it uses big data methods, with profit, in one part of an overall prediction process: cluster analysis of the exact relations between demographics and voting behavior.[37] MRP modeling is not a magic bullet—its forecasts are still fallible. But again, the impact of big data is not all-or-nothing; rather, it brings incremental improvements. There are other examples of that in election forecasting, such as pioneering work that uses big data methods to improve impact response analysis, that is, to infer implications for voting behavior from the speed and way respondents answer questions.

Another lesson from case studies is the necessity of background theoretical knowledge. Such knowledge is often an essential guide to choosing the Pietschian correct vocabulary, and it is often essential, too, for correcting fragility—when that is feasible—and for guiding extrapolation.[38] That is, theoretical knowledge is beneficial even just for the narrow goal of better prediction. Sabina Leonelli emphasizes this point for systems biology.[39] With the auctions, it was theory that informed the experiments that gathered the relevant data. MRP election modeling requires background theory too, because it works well only if it considers a well-chosen range of demographic variables. This shows up when MRP modeling is applied to a new country. Local experts are required to advise on the choice of parameters: in some countries, for example, voting is strongly influenced by occupation, whereas in others it is not, and it varies country by country how best to divide by age groups.[40] Generally, it is a commonplace among practitioners that background theory and knowledge of data-generating mechanisms are essential for handling data errors effectively.[41] For these and other reasons, as many have argued, the "death of theory" is greatly exaggerated.[42]

A note of caution: although theory is indispensable, at the same time, theory alone does not predict successfully. All the examples confirm that. The weather forecasting models require many ad hoc adjustments that go beyond, or even contradict, basic theory, as discussed in Chapter 6; models of elections based on fundamentals are outpredicted by opinion polling; theory-based forecasts

---

[37] Personal communication with the polling company YouGov.
[38] See also Knüsel et al. (2019, 199).
[39] Leonelli (2016).
[40] Personal communication with the polling company YouGov.
[41] Foster et al. (2017, 180).
[42] Hansen and Quinon (2023); Pietsch (2021); Boon (2020); Calude and Longo (2017); Leonelli (2016).

of GDP fare no better than other methods; and the spectrum auction design was not derivable from game-theoretical models but rather used those models only as heuristic starting points. Recent work suggests that machine learning methods likewise predict best by avoiding theory-driven models.[43] In our case studies, more data do not counter this antitheory trend; if anything, they exacerbate it.

What of big data practitioners? Their analyses overlap with Pietsch's. Correct choice of variables is recognized as key for machine learning, for example. But practitioners emphasize things differently. Perhaps the biggest issue for them is data quality, which is a catch-all for a range of more specific issues, such as whether data are representative, whether there are measurement errors, and whether the data capture what we want them to—data are often repurposed because they are the products of instruments and methods not developed with data scientists in mind.[44] Such concerns are captured by the Pietsch conditions at best only implicitly.

In our examples, though, while data quality issues such as the reliability of polling evidence and weather measurements are concerns, they are not the main problem. A bigger problem is fragility. Benedikt Knüsel and colleagues are one group of practitioners who agree, as they take what they call "constancy" in the data to be, in practice, the most important condition to satisfy.[45]

Overall, big data methods advance field prediction significantly. They enable the best possible use to be made of a given body of data, and they promise further advances with new techniques and more data. But they are not a panacea. Although lack of data is one constraint on predictive success, often the binding constraint is neither lack of data nor inefficient use of them. Instead, it is fragility. And big data is no magic bullet for that.

---

[43] Mullainathan and Spiess (2017).
[44] Foster et al. (2017, 276–85); Japec et al. (2015, 848–50).
[45] Knüsel et al. (2019).

10

Fragility and the COVID-19 pandemic

10.1 Fragility and COVID

What kind of epidemiological modeling works well, and what kind does not?[1]
The COVID-19 pandemic stimulated a vast amount of epidemiological work.
I look at two high-profile examples, each of which influenced policy in several countries. The first example is the CovidSim model in "Report 9" by Neil
Ferguson and colleagues at the Imperial College COVID-19 Response Team,
the dire projections of which were a large factor in persuading the UK government to impose a national lockdown in March 2020.[2] The second example
is a paper from December 2020 by Erik Volz and colleagues, also at Imperial
College London, which was the first to estimate the greater transmissibility of
the Alpha variant.[3]

The lesson of these examples is that the relevant models do not apply reliably. Epidemiology is not like Newtonian physics. The reason is that key relations are fragile, so extensive case-specific investigation is needed each time to
know which—if any—model applies. As we will see, this is why the Volz paper
succeeds but the Ferguson paper fails.

The distinction between Case-Worker and Stability-Theorist cross-cuts
previous distinctions in the epidemiological literature, such as that between
model types (agent based, compartmental, curve fitting) and that between
interpreting models causally and noncausally.[4] Each model type, and both
causal and noncausal models, may be developed and used in accordance with
either Case-Worker or Stability-Theorist.

---

[1] This chapter is adapted with permission from Northcott (2022a).
[2] Ferguson et al. (2020).
[3] Volz et al. (2020).
[4] Fuller (2021).

Science for a Fragile World. Robert Northcott, Oxford University Press. © Robert Northcott 2025.
DOI: 10.1093/9780191944352.003.0010

10.2 The CovidSim model

March 2020 was a key moment in the UK's pandemic response. Infections were rising rapidly, and momentous decisions had to be made in a hurry: how many would die given different policy interventions, or given no interventions at all? Different models competed to advise. An especially influential one was developed by Neil Ferguson and colleagues at Imperial College, London. Their model was one of the main factors that persuaded the UK government to reverse its previous policy and to impose a national lockdown.[5]

Report 9 does not itself develop a new model; rather, it adapts a transmission model originally developed by the same team for outbreaks of influenza. This adapted model—CovidSim—does not just analyze the dynamics of population-level aggregates, in the manner of the classic SIR (Susceptible–Infectious–Recovered) model. It is more ambitious. CovidSim is a hybrid of agent based (i.e., it models the actions of individual agents) and compartmental (i.e., it models interactions between different groups in the population). It models interactions between individuals within the home, at school, at work, and in the community. Actual data are used to estimate values for parameters such as age and household size, class size and staff–student ratios, and workplace size and commuting distances. COVID-specific values are added for the virus' incubation period, reproduction number (R), and infection fatality rate.

Using CovidSim, Report 9 analyzes two kinds of policy intervention. The first it labels *Suppression*: use restrictive measures to drive the virus' R below 1 and so reduce case numbers, in the hope that vaccines or treatments with high efficacy are eventually developed. Until vaccines or treatments arrive, restrictive measures must be reimposed periodically because infections will rise again whenever measures are eased. The report labels the second kind of intervention *Mitigation*: less strong restrictive measures, which reduce R but not all the way to below 1. The aim is to spread out the number of cases over time (compared to with no intervention) so that at no point are health services overwhelmed, in the hope that immunity in those previously infected builds up gradually in the population and eventually leads case numbers to decline. This eventual decline would happen even without vaccines. Report 9 concludes, based on its simulations, that any feasible version of Mitigation would leave national intensive care unit (ICU) capacity overwhelmed by a factor of at least 10 and would inevitably lead to many deaths. Therefore, Suppression is

[5] Grey and MacAskill (2020); (Ford 2020); Conn et al. (2020).

preferred. Acceptance of this recommendation by the UK government was the paper's major policy impact.

In which methodological camp does the CovidSim model belong, Stability-Theorist or Case-Worker? On one hand, it aims to be more sensitive to local circumstances (via agent-based modeling) than are traditional epidemiological models such as SIR, and accordingly, it has many more parameters. But overall, it clearly belongs in the Stability-Theorist category. It is intended to apply to many infectious disease epidemics. Each time, predictions are generated by the same underlying model with the same structural relations and parameters, just with different values for those parameters.

I will be critical. But I want to acknowledge that the CovidSim model—an intricate piece of work—was produced at speed, at a moment of national crisis and was a well-intended attempt to guide policy.

10.3 Criticisms of the CovidSim model

The criticisms below are structured. The first and most important criticism is lack of empirical confirmation. I then consider the defense that the CovidSim model may gain warrant even without empirical confirmation. I reject this defense because, first, relevant parameter estimates are inaccurate and, second, important relations are omitted. After thus establishing that the CovidSim model lacks warrant, I diagnose the underlying reason why: it treats as stable, relations that are fragile. This undercuts hope that the model can be fruitfully augmented.

10.3.1 Lack of empirical confirmation

The most important problem is lack of empirical confirmation. As many have noted, assessing this needs care because the CovidSim model gives only projections.[6] In other words, its predictions are conditional: how many infections, and how much pressure on health services, *would* there be given various policy combinations? Actual policy did not replicate any of these combinations exactly. Are criticisms of the model's poor predictive record therefore unfair?

In reply, two points. First, if a model cannot be tested empirically, that is a bad thing, not a good thing. Such a model cannot receive empirical

---

[6] Fuller (2021); Schroeder (2021).

feedback or refinement, which blocks off the main route to scientific progress. Confirmation will remain absent forever.

Second, in fact, the CovidSim model can be tested, at least somewhat. For policy scenarios like the actual policy mix, were its projections approximately right? No, they were not—or at least, I think that is the answer that emerges from scholarly to and fro on this question.[7] Eric Winsberg, Jason Brennan, and Chris Surprenant argue that the model greatly overestimated deaths and ICU usage in Sweden and Florida.[8] Other projections of Report 9 were also inaccurate, such as how quickly deaths would accumulate, how long policy measures should be maintained to avoid ICU capacity becoming overwhelmed, and how long policy measures could be eased before case numbers rose too high again. The CovidSim model clearly applied poorly to places such as Africa, where infection numbers were far lower than predicted.[9] Other failures cement the same negative verdict. The Scientific Advisory Group for Emergencies, the official scientific advisory body to the UK government, informed by Report 9, commented on March 20, 2020: "it is very likely that we will see ICU capacity in London breached by the end of the month, even if additional measures are put in place today."[10] But London's ICU capacity was not breached. And Ferguson himself, in late March 2020, after lockdown had been imposed, and again informed by the CovidSim model, predicted to Parliament that total UK deaths would top out at about 20,000, which of course sadly proved a huge underestimate. There is also no track record of predictive success in other epidemics to confirm the model independently—although such success elsewhere would be relevant anyway only if (implausibly) relations were stable enough for warrant to carry over from, say, an influenza epidemic to the COVID one.

Prediction in epidemiology is challenging generally,[11] so this poor performance should not be jeered at. The point is only to note its epistemic implications.

Does the CovidSim model nonetheless capture accurately some of the causal relations involved, and so still explain partially? That is possible. But such a claim needs warrant, and we do not have it.

---

[7] Winsberg et al. (2020, 2021); van Basshuysen and White (2021a, 2021b); White et al. (2022).
[8] Winsberg et al. (2021).
[9] Broadbent and Smart (2020).
[10] SPI-M-O (2020).
[11] Broadbent (2013).

10.3.2 False assumptions

The CovidSim model is idealized: many of its assumptions are false. Does this matter? Not necessarily. A lot of recent philosophy of science has addressed this question, and the consensus is that, roughly, idealized models can explain when their falsity does not matter—when their idealizations are true enough for the purposes at hand. The sting in the tail, though, is that empirical endorsement is still needed at some point. A Newtonian model of gravity, for example, makes false assumptions, yet because it predicts accurately, and because for many purposes its assumptions are "approximately" true, it is rightly considered explanatory. But the predictive endorsement is essential, and that is just what the CovidSim model lacks.

There is an alternative defense: that a model's assumptions are so compelling that empirical confirmation of the model's results is not essential. Even if not confirmed empirically, on this view, a model retains epistemic force as a guide for interventions, and it explains partially. But for the CovidSim model, this defense is unconvincing. The model's assumptions are not compelling, as we will see now.

A few of the CovidSim model's parameters can be estimated relatively straightforwardly. Already, difficulties arise. Two of these "easier" parameters are the virus' reproduction number and its infection fatality rate. The model estimates these only to be within a range: 2.0 to 2.6 for R, and 0.25 to 1% for the infection fatality rate. But subsequent work suggests that, in both cases, the true values were outside these ranges: in a society such as the UK, without social distancing measures, a central estimate of R (i.e., R_0 for the original variant) is 3.0, while a central estimate of the infection fatality rate in the UK's first wave is 1.1%, albeit new treatments lowered the rate later.[12] Further, the estimates of Report 9 are meant to cover all of the policy scenarios that it analyzes, but they do not. The scenario of zero social policy interventions, for example, produced, for the UK, the headline-catching projection of 510,000 deaths. Yet with zero policy interventions, ICU capacity would likely have been swamped—by a factor of 30 according to the report's own estimates—so that only a tiny proportion of seriously ill COVID-19 patients could have received the best treatment. In that circumstance, given that the hospitalization rate in the UK of COVID-19 patients in the first wave was about 4% to 5%, the infection fatality rate would likely have been well above the report's estimate of 1.1%.

---

[12] Billah et al. (2020) estimate R, and Brazeau et al. (2020) estimate the infection fatality rate.

And to repeat, these difficulties concern reproduction number and infection fatality rate, two of the easier parameters to estimate. Other parameters require a lot more work, and some can be estimated only with educated guesses. If symptomatic COVID-19 patients are required to stay at home, then how many days at home? The report estimates seven days. What would be the effect on patients' contacts? The report estimates that nonhousehold contacts would decline by 75%, while within-household contacts remained unchanged. How many households would comply with this policy? The report estimates that 70% would. Values for other parameters are educated guesses too: that a symptomatic patient's household members would comply with voluntary home quarantine 50% of the time and that 75% of those over 70 would comply with social distancing, reducing their contact by 50% in workplaces and by 75% in the community, while increasing it by 25% within households. Educated guesses are made about the impacts of social distancing and of schools and universities closing. How long these various measures last is also relevant; the report estimates that each would last three months, except for social distancing of those over 70, which would last four months.

Inevitably, many of these educated guesses turned out to be inaccurate. Notably, policy measures, such as two-meter social distancing, mandatory mask-wearing in indoor public spaces, and closure of nightclubs, lasted in the UK for almost a year and a half—far longer than the three months assumed in Report 9.

These details show that Report 9's assumptions are not so compelling as to be self-warranting. But perhaps its projections are not sensitive to precise parameter estimates, and all we need is those estimates to be roughly correct. If so, then we need to know how roughly. The report does give one sensitivity analysis—but of very limited scope. This analysis shows that the main policy recommendation, in favor of the Suppression policy over the Mitigation one, is not sensitive to the precise value of R, the infection fatality rate, or to what number of cases would trigger policy interventions. Or at least, the analysis demonstrates the insensitivity result for values of these parameters that the paper considers plausible, such as between 2.0 and 2.6 for R—but remember, it is likely that R's actual value was above this range. And there are many other parameters in the model. How sensitive is the report's main policy recommendation to those, and how sensitive are its quantitative projections? Given that the model has hundreds of parameters, it is doubtful that a full sensitivity analysis for the main conclusions is even feasible. Certainly, none is given. As with several other shortcomings, Report 9 does acknowledge this problem but offers no solution.

Two projections of the report do clearly fail a sensitivity test. The first is the estimate of 510,000 UK deaths in the absence of any policy intervention. This projection assumes an infection fatality rate of about 1%. But as noted above, if ICUs and hospitals were overwhelmed, the infection fatality rate would likely be well above 1%, which would increase the total number of deaths in direct proportion—an infection fatality rate of 2%, for example, would raise the figure for deaths to over 1 million. The second projection is that, even in the best Mitigation scenario, 250,000 would die in the UK. This projection is central to the main policy advice of Report 9. But note how sensitive it is to a single assumption, by the report's own admission (italics added):

> In the UK, this conclusion has only been reached in the last few days, with the refinement of estimates of likely ICU demand due to COVID-19 based on experience in Italy and the UK (*previous planning estimates assumed half the demand now estimated*) and with the NHS providing increasing certainty around the limits of hospital surge capacity.[13]

That is, when the estimate of ICU demand was updated, the estimated number of deaths under Mitigation thereby doubled—almost overnight.

More generally, later work suggests that the projections of Report 9 are highly sensitive not only to its estimates of parameter values but also to omitted factors (see shortly) and to uncertainty about what actual conditions were.[14] This reinforces the point that the report's own sensitivity analyses are not enough. True, time was of the essence, so only limited sensitivity analysis was feasible. But this practical constraint does not alleviate the epistemic problem.

10.3.3 Omitted relations

Incorrect parameter estimates are not the only problem. In addition, in Report 9, many relations are omitted. Consider just two mentioned by the report itself. First, the degree of social contact is influenced by people's spontaneous behavioral responses: high case numbers lead to more social distancing spontaneously and, conversely, low numbers lead to less distancing. Second, if schools are closed, this reduces health service capacity because some health workers who are also parents are forced to stay at home. Each of these omitted relations

---

[13] Ferguson et al. (2020, 16).
[14] Edeling et al. (2020); Winsberg et al. (2021).

implies that parameter values assumed by the model to be constant—levels of social contact and health service capacity—are, in fact, functions of other variables in the model. Again, an assumption—here, constant values for these parameters—is clearly not so compelling as to warrant itself.

10.3.4 No jam tomorrow: instability

The above range of failures is true of rival models, too. None is predictively endorsed, and all make false assumptions and omit relevant relations. Is there, nonetheless, still hope for Stability-Theorist? Perhaps the pandemic is characterized by stability plus noise. If so, then even though no current master model has acquired empirical warrant, a future one might do better.

But alas, disappointment with Stability-Theorist here is likely to be permanent. Why? Because of fragility. We have just noted one source of it: omission of some relations in Report 9 makes other, unomitted relations—that is, relations that are in the CovidSim model—fragile. The general picture is that many key relations vary unpredictably across time, place, and virus type, so Stability-Theorist is unsuitable.

Consider border controls—a key policy tool for many countries. How do border controls impact infection rates? The answer varies with the idiosyncrasies of a country's borders, trade flows, location on transit routes, number of residents with ties abroad, and number of border personnel and hotels, not to mention the nature of the relevant virus and disease. When should border controls be triggered, for how long maintained, how long should quarantine periods be, and what exceptions are allowed—some countries, some freight, airline personnel? Correct answers cannot be predicted reliably. Instead, they have had to be worked out via trial and error, case by case, country by country.

Similar remarks apply to test and trace programs. Compare the programs needed for COVID-19 with those needed for SARS in 2003, or, within the UK, compare outsourced private operations with those run through local councils' public health officers. Similar remarks apply to public health messaging too. Compare, say, the very different media and political environments in Singapore, the UK, and the United States. Each time, local investigation is needed to work out a policy's impact.

Report 9 did not model any of border controls, testing programs, or public messaging. Many relations that it did model are likely just as fragile: the key factor of how long a policy is sustained is unpredictable because popular resistance to lockdown measures has varied unpredictably across countries, over time, and more for some measures than for others; the value of R is not a constant but rather is a function of environments, behaviors, and political decisions; and so too, is the time between infection and transmission. Report 9 states that "stopping mass gatherings is predicted to have relatively little impact."[15] But while this is true for influenza, community transmission of COVID-19 seems to have been powered disproportionately by large "super-spreader" events, so stopping mass gatherings in fact does have a large impact. The relevant relations are unstable across influenza and COVID epidemics. More generally, Report 9 omits social network effects and therefore misses potential interventions targeted at hub actors.[16]

This underlying instability is not just because the COVID virus was new to science. If anything, new knowledge causes some relations here to become less, not more, stable. The efficacy of government restrictions, for example, changes once authorities learn how to implement them better, once vaccines become available, and once public obedience wanes.

Compare the CovidSim model to a paradigm of stability: a Newtonian model of gravity. As discussed in Chapter 4, although a moon's position and velocity vary continuously, this is not a problem for the Newtonian model because the model tells us not just the effects of that variation but also when to expect it. There is no "surprise" variation in gravity's influence *that requires knowledge from beyond the model* to predict. But the same is not true of the pandemic. Consider the relations between R and other components of CovidSim, for example. These relations are fragile. They vary with contextual factors such as virus characteristics, local history, and local politics, so now we do require knowledge from beyond the model itself.

10.4 Verdict

Stability-Theorist's most important virtue is efficiency: a master model is a shortcut to successful predictions, interventions, and explanations. But with the CovidSim model, this efficiency gain no longer exists. The model is over-ambitious: too many relations in it are fragile, and Stability-Theorist is misconceived from the start.

---

[15] Ferguson et al. (2020, 8).
[16] Manzo (2020).

Where does that leave us? In my view, the CovidSim model has no epistemic force at all. Its predictions have a track record of inaccuracy, and the model is not confirmed in any other way. It omits many important relations, many of the relations it does include are likely misspecified, and many parameter values are estimated inaccurately. No sensitivity analysis reassures us that these errors are not fatal. According to no philosophical theory of explanation does such a model explain.[17] We have no warrant to think the model has captured the true causal structure even in part, and so no warrant to trust it as a guide to interventions.

The literature on idealized models does offer two other potential defenses, but neither of them helps here. The first defense is that, even when they lack empirical warrant, idealized models may give "how-possibly explanations": accounts of how things work in a hypothetical world in which the model's assumptions are true. The hope is that how-possibly explanations shed light on the actual world indirectly because they show how things could be or could have been. But the CovidSim model explicitly aims to model the actual world—to its credit, given that its goal is to advise policy. The second defense of idealized models is that, even when they do not themselves predict or explain, still they are useful heuristically because they direct our attention to important factors otherwise overlooked.[18] But Report 9 actively turns our attention away from key omitted factors. And it also guides us away from the methods that, as we will see in the next two sections, do deliver.

In a crisis, speed matters—a constraint on optimal methodology is the context in which a question is investigated. Stability-Theorist does offer a pragmatic advantage: a model is available "off the shelf." The CovidSim model was certainly adapted quickly from its influenza origins. But speed is not the only thing. If a model lacks epistemic force, that defect trumps speed.

The CovidSim model influenced policy when it should not have. Because it inspired a lockdown, perhaps it had highly beneficial consequences. If so, that was by luck. A judgment that the lockdown was indeed beneficial should be based not on the CovidSim model but rather on accumulated experience from many countries, assessed by other methods.

---

[17] Northcott and Alexandrova (2013).
[18] Alexandrova (2008); Alexandrova and Northcott (2009).

10.5 Example of Case-Worker: transmissibility of the Alpha variant

The pandemic has provided many examples of Case-Worker in action. One is the paper that first established the higher transmissibility of the Alpha variant, written in December 2020 by Erik Volz and colleagues at Imperial College London and elsewhere.[19] Unlike those of Report 9, this paper's conclusions are compelling.

One of the coauthors of the Volz paper is Neil Ferguson, and several of the other coauthors, including Erik Volz, are also coauthors of Report 9. (Some are coauthors of the later "Report 34" that estimated the infection fatality rate, as well.[20]) Nevertheless, as will become clear, I think the difference in epistemic force between the Volz paper and Report 9 is stark.

The Volz paper has narrower scope than Report 9. It estimates the transmissibility of the then-new B.1.1.7 COVID "variant of concern," that is, what the World Health Organization later designated the Alpha variant. Its data are all from England between October and early December 2020. Experiments were not feasible, so the paper conducts observational studies that combine epidemiological and genetic evidence. It pursues five independent lines of analysis, each of which turns out to concur on the same conclusion: Alpha is more transmissible than the original variant, to the extent that it increases the virus' R (in England, in this period) by between 0.4 and 0.7. These five lines of analysis are:

1. The time and location of almost 2,000 Alpha cases from random population samples were tracked, along with almost 50,000 non-Alpha cases. This revealed the growing prevalence of Alpha relative to the original variant. To infer the growth difference per generation, a simple model was used of two variants with different reproduction numbers. This model required the paper to estimate the virus' generation time (6.5 days) and the original variant's R at that time (1.0).

2. The growing prevalence of a genetic feature associated with Alpha—the absence of the so-called S-gene—was traced, based on data from national positive COVID test results. This was possible because, conveniently, almost a third of positive test samples in November and December 2020 recorded the presence or absence of the S-gene. The strength of

---

[19] Volz et al. (2020).
[20] Brazeau et al. (2020).

association between Alpha and absence of the S-gene itself varied over time (because some non-Alpha variants also lack the S-gene), and this extra variation had to be modeled as a function of the date and area of the test sample. Overall, it turned out that the spread of Alpha inferred in 1 and 2 matched closely.

3. The pattern of geographical expansion, as opposed to national prevalence, of absence of the S-gene was tracked, again using data from national positive COVID test results, and again inferring Alpha prevalence by means of the intermediary model mentioned in 2. The result was consistent with a greater transmissibility for Alpha, to a similar degree as calculated in 1 and 2.

4. A positive correlation was established between the estimated prevalence of Alpha and independently derived estimates of the overall COVID-19 reproduction number at different times and places. The paper ran a series of statistical regressions, with different measures of Alpha prevalence, different subdivisions of areas, and both frequentist and Bayesian estimation techniques. It derived quantitative estimates of the increase in R associated with Alpha. These estimates roughly agreed with those derived from the other lines of analysis.

5. A semi-mechanistic genetic model was fitted to the case numbers for Alpha and the original variant, to derive from many separate regressions further independent estimates of R for each. These new estimates again roughly agreed with those derived from the other lines of analysis.

The Volz paper was also able to rule out rival explanations for Alpha's greater prevalence.[21]

Overall, the Volz paper is excellent work. How does it succeed? It uses models that are relatively simple, and it uses them in a suitably contextual way. This allows it to be confident that its models apply, as the details above show. In the paper's words: "we focused on relatively simple, data-driven analyses using parsimonious models making parsimonious assumptions, rather than, for instance, attempting to model the long-term transmission dynamics of [the Alpha and original variants] more mechanistically."[22] In other words, it did not follow the CovidSim example—and wisely not. Narrower scope in return for empirical confirmation is a good trade.

[21] See Northcott (2022a) for details.
[22] Volz et al. (2020, 18).

To illustrate, consider the model in the paper's first line of analysis. It models how the relative prevalence of two variants with different reproduction numbers changes over time, and it requires only two parameters to be estimated: the virus' incubation period and R for the original variant. This model is disanalogous to the CovidSim model in several other respects. Most importantly, it has a history of empirical success. Here, the model captures the dynamics of virus reproduction when social factors do not interfere—and in this case, experience gives us confidence that social factors indeed did not interfere significantly. Social factors enter the model indirectly via the value of R, but for the relevant times and places, there were good independent estimates of R's value and good reason to think that that value was roughly constant. The virus' incubation period, too, was independently known. And straightforward sensitivity analysis shows that the results are not unduly hostage to remaining small uncertainties.

For these reasons, confidence in this model is warranted here, unlike for the CovidSim model. Similar remarks apply to other models used in the Volz paper. And similar remarks apply, too, to many other excellent studies carried out during the pandemic. What matters each time is that a model is developed and used in accordance with Case-Worker rather than Stability-Theorist, and empirical confirmation is crucial for that.

The important difference is not that in the Volz paper models were used for retrospective estimation while in Report 9 they were used for forward-looking projection. Rather, the important difference is that the models in Volz have empirical warrant whereas the CovidSim model does not.

There is a caveat, though. In accordance with Case-Worker, predictions are warranted only when we have good reason to believe that the models behind them still apply—and this implies narrow scope. The Volz paper acknowledges this: "these estimates of transmission advantage apply to a period where high levels of social distancing were in place ... extrapolation to other transmission contexts ... requires caution."[23] This is wise. Extrapolation requires caution whenever relations are fragile. Here, reproduction numbers are known to be sensitive to many environmental changes, and indeed, Alpha's transmissibility advantage did not stay the same after the period of the study.[24] Some are skeptical that it was constant even within this period.[25] Others are more confident. Either way, confidence should be in proportion to empirical confirmation,

---

[23] Volz et al. (2020, 17).
[24] Lemoine (2021a).
[25] Lemoine (2021b).

which here means in proportion to the extent that Case-Worker was followed. Further, there was a pattern throughout the pandemic of new variants enjoying large initial transmissibility advantages that then diminished sharply, which again suggests caution when extrapolating initial calculations.[26]

10.6 Informal methods

Sometimes, the only methods available are informal ones. These amount to causal reasoning based on evidence, including quantitative evidence, but done in the manner of careful historians or qualitative social scientists. When relevant relations are fragile, as they were in the pandemic, then if such work is done in accordance with Case-Worker, it should be endorsed.

The "models" in these cases might be no more than simple causal claims, such as "border controls reduce Covid cases." Usually, the claims are more detailed: "if border controls are organized in manner X at stage Y of the pandemic, then in countries with a high throughput of travelers the controls reduce Covid cases by Z." How to confirm such claims? By approximating natural experiments as best we can or by single-case causal inferences that rely on background knowledge to evaluate the implicit counterfactuals. Karen Ann Grépin and coauthors, for example, review early work on border controls to infer that at the beginning of the COVID-19 pandemic, travel restrictions around Wuhan reduced the spread of cases internationally—contrary to the experience of influenza outbreaks.[27] Even though their methods are informal, they still obtain empirical warrant. And it is by informal methods that best practice about border controls has been established and shared, such as operational details of hotel quarantine, how controls should be tailored to a traveler's country of origin, or how their impact varies depending on current case numbers and on whether those numbers are rising or falling. The key epistemic criterion is the same for all models alike, formal or informal: empirical confirmation.

Similar remarks apply widely. Consider the vexed issue of lockdowns. There has been a huge amount of commentary about them, of course, which I will not review; I mention lockdowns only to point out several ways in which fragility bears on them. What is their impact on infection numbers, economic output, and mental health? First, there is no univocal answer. Impact varies with the exact lockdown regulations imposed, by whom, on what community,

---

[26] Lemoine (2021b).
[27] Grépin et al. (2021).

and at what stage of the pandemic. During a recession, in winter or summer, in an urbanized or rural country, in a rich or poor one? In a country with lots of gig workers, with lots of multigenerational households, with a history of suspicion of government? On a population that is healthy, with low levels of comorbidities, with access to high ICU capacity and extensive primary care? Even within a single country—the UK—the impact of the same lockdown regulations, at the same stage of the pandemic, differed greatly across regions and sectors. Fragility abounds. Second, Case-Worker is required. The impact of lockdowns should not be assessed via a single master model such as CovidSim. Third, is there nonetheless some rule of thumb that is useful widely, perhaps that lockdowns' effects are overrated or underrated? Perhaps, perhaps not. The question can be answered only by close empirical analysis, in the manner of the Volz paper. When there are nuggets of stability to be found, Case-Worker is the best way to find them—a maxim of wide significance, as we saw in Chapter 4.

Because countries vary so much in relevant ways, it is not helpful to assess the impact of lockdown policy by running a simple statistical regression across countries. For the same reason, a hypothetical randomized trial would tell us little. As I discuss in Chapter 6, methods such as regressions and trials are very effective—but only when relations are stable. That is why they work well for assessing vaccines but not for assessing lockdowns.

Similar remarks apply to many other pandemic policy questions: how much to test, what mask-wearing and social distancing to require, whether to close schools and universities, how to regulate those asked to self-isolate, how to contact-trace, and how to allocate medical equipment. What can be learned by comparing the experiences of different countries? Which local details matter, and which do not? Informal methods are usually the only way to find out.[28]

How could we have used informal methods to tackle the original task of Report 9, that is, to project the number of infections under different policy scenarios? By going Case-Worker. We would have needed to assess COVID-19 developments in China, Italy, and other countries, experience of previous epidemics, and background knowledge of local health systems and political cultures. Even very early in the pandemic, there was already plenty of such evidence.[29] Still, most likely, only very rough projections could have been justified initially. If so, it is better to accept that than to imagine that more precise projections delivered by Stability-Theorist modeling are trustworthy. If a model

[28] Han et al. (2020).
[29] Lipsitch (2020).

is unwarranted, it is not probative—and that does not change just because an epistemic situation is difficult and there are few alternatives.

There is continuity between informal methods and the methods of the Volz paper. How formal a model is, and how formal the techniques are by which to test whether a model applies, may vary. But both informal methods and the Volz paper share the same Case-Worker strategy, and that is what matters.

10.7 Last word

The Stability-Theorist strategy does not work for the epidemiology of the COVID-19 pandemic. No candidate master model boasts the needed empirical success, and likely none will, because the relations involved are fragile. A different strategy is required. The same conclusion applies whenever relations are fragile, which likely means for pandemics generally.

This conclusion is not of mere ivory-tower interest. To pursue Stability-Theorist is harmful when it diverts resources away from a superior Case-Worker alternative. Facing a policy emergency in March 2020, the initial choice was between models such as CovidSim versus informal methods. The UK government's Scientific Advisory Group for Emergencies was criticized for being initially top-heavy with mathematical modelers rather than empirical field scientists.[30] By the time this began to be (slightly) rectified, fateful policy mistakes had already been made: relative neglect of on-the-ground experience from other countries and from practitioners at home is widely alleged to have slowed the provision of protective equipment to health workers and to have slowed the roll-out of a testing system.

We know a lot more now, so both projections and policy responses are far better grounded than they were. But this welcome progress has not come from grand, one-size-fits-all models. Rather, it has come from a huge accumulation of knowledge gained by informal methods and by modeling that is empirically confirmed. Case-Worker was, and is, needed.

---

[30] Ford (2020); Costello (2020).

11

Conclusion: expertise in a fragile world

11.1 Summary of the book

There is a core divide between two methodological strategies. The *Stability-Theorist* strategy advocates that we develop theory under the assumption that relations hold reliably. It licenses us to extend and elaborate a theory freely, confident that the relations in our extensions and elaborations will continue to hold. Empirical confirmation may be left on autopilot. Stability-Theorist's great advantage is efficiency: we do not need to investigate each time whether our theory holds. Instead, to apply the theory is just a matter, roughly, of tweaking parameter values and auxiliary assumptions case by case, or in Mill's phrase, adding in "disturbing causes." Often, Stability-Theorist allows for a single, master theory for a domain, such as Newtonian mechanics. We may funnel resources into this master theory and refine it for endless new applications, rightly confident that it will retain empirical purchase all the while. This efficiency lies behind many of science's most famous achievements.

The *Case-Worker* strategy, by contrast, accepts that we cannot tell in advance which (if any) of our theories apply because none of them operates reliably enough. An explanation established for one case does not reliably extrapolate to another; its warrant is narrow scope. Supplementary, historian-like investigation of the details of a case is required each time. There is no single master theory. The efficiency promised by Stability-Theorist is now an illusion: theories developed without continuous empirical refinement risk degenerating into houses of cards. Instead, theory must be developed via case studies and frequent empirical applications—no more leaving on autopilot. When tackling a real-world target, we choose each time from a toolbox of theories and other less formal resources, guided by whatever turns out to apply well, not knowing the answer in advance.

Which strategy to choose? It all depends on how reliably an explanans relation holds. If it holds reliably enough, then Stability-Theorist; if not, Case-Worker. Exactly how much unreliability tips the balance varies case by case

Science for a Fragile World. Robert Northcott, Oxford University Press. © Robert Northcott 2025.
DOI: 10.1093/9780191944352.003.0011

and purpose by purpose. But whenever that threshold is breached, a relation is usefully labeled *fragile*.

How often, in fact, are relations of interest fragile? A lot, including many of the relations we care about most. Fragility's consequences need to be reckoned with.

How can we tell if a relation is fragile? There is no mechanical algorithm. But in practice, as illustrated by numerous case studies, whether a relation is fragile is usually obvious soon enough. There is little danger of somehow missing out on stability by settling prematurely for Case-Worker; the danger is overwhelmingly the other way around—of pursuing Stability-Theorist inadvisably. And when facing uncertainty, the best way to discover whatever oases of stability do exist is, ironically, via Case-Worker.

Fragility bears on familiar scientific methods. It suggests the value but also the limitations of laboratory, statistical, and big data methods; it endorses qualitative methods; but it tells against theory monism. It also bears on philosophy of science. When relations are fragile, only some forms of scientific realism are plausible, and instrumentalism never is. Mechanistic explanation is endorsed, but mechanisms must be developed in the right way. Reflexivity is overrated as a threat to social science. And scientific progress, although all around us, is inevitably messy and piecemeal, in some ways not as obvious as when relations are stable.

These lessons carry bite. Throughout the book, we have seen in detail the baleful consequences of ignoring them. The examples of bad practice share a common error: Stability-Theorist is pursued when that is a mistake because explanans relations are fragile. More positively, often these same examples also showcase Case-Worker achieving much in difficult epistemic circumstances, such as Ashworth's explanations of the World War One truces, della Porta's theories of political violence, the modeling of invasive pine species, and the estimation of the Alpha variant's transmissibility. These are exemplars. Case-Worker works: science can succeed even in the face of fragility.

All of this tells against common slogans: that science aims at true theories, scientific progress is constituted by new theories, good theories may be viewed realistically, science is about discovering stable mechanisms, science is about discovering stable capacities, case studies are mere anecdotes, theories must be mathematically precise, only quantitative methods deliver causal inference, statistical work should aim to test theories, a model's false assumptions do not matter, big data will transcend current scientific methods, and social science needs a unified theoretical framework. In a fragile world, none of these are true.

11.2 Example of the core divide: two kinds of insurance underwriting

A final, short example encapsulates the themes of this book.[1] It illustrates again how the core methodological dichotomy between Stability-Theorist and Case-Worker arises spontaneously, corresponding to whether key relations hold predictably or not.

To set premiums, insurance companies must assess risks, which requires modeling the processes that generate those risks. The relations in these processes can be either stable or fragile. In response, in just the way we should expect, two distinct modeling strategies are employed, corresponding to Stability-Theorist and Case-Worker.

Consider, first, personal lines of business such as car insurance or house insurance. For these, insurance companies use "flow underwriting," sometimes also called "fast flow" or "e-traded" underwriting. Statistical data are used to establish what influences how many claims are made: factors such as the type of car, the age and experience of a driver, or the construction type of a house's roof. Premiums are calculated accordingly. The analysis is large-N, and like all such analyses, it works only because the key relations are stable: the relation between age of the driver and expected number of claims, for example, holds reliably. The same statistical model is used year after year, modulo tweaks to parameter values—Stability-Theorist in action. Although new factors are added to the model occasionally, it is tinkering at the edges.

Consider, second, some commercial lines of business such as a multinational firm insuring against complex energy infrastructure risks, which might include such things as refineries, oil rigs, and offshore wind farms, or an airline insuring its entire fleet of planes. These are different from car or house insurance. The size of policies and the exact perils insured against vary a lot, and so in high-value cases "case underwriting" is used instead of flow underwriting. This requires a different kind of modeling. Usually, there is great uncertainty about how best to interpret and process the data, and judgment calls are ubiquitous. The actuary does still build a predictive model but now that model is bespoke to each case, and to get it right requires supplementary research each time. The strategy is Case-Worker.

This example does not just illustrate the core divide between Stability-Theorist and Case-Worker. It illustrates other resonant features too.

---

[1] I owe this example to Simon Pollack (2022), a student at Birkbeck with a distinguished career as a pricing actuary.

Case-Worker is required for case underwriting even though calculations of the bespoke estimates may depend on other relations that are stable. For example, the calculation of marine and aviation risks might incorporate underlying accident rates that are stable, even while exactly how those rates combine and bear on the premium is unique to the insurance case at hand. "One more heave" is hopeless: it would be absurd and inefficient to search for some repeated relation to analyze statistically. The relation of interest is between a specific set-up and the expected size and chance of a claim that results from it, and this set-up might be unique. Only Case-Worker will do.

The dichotomy between flow and case underwriting exactly tracks that between Stability-Theorist and Case-Worker. This vindication carries authority. Like the instrumentalism of weather forecasting, it results not from philosophical fancy but from commercial pressures that eliminate any modeling that is wrong-headed.

11.3 What expertise is not

Turn now to expertise. Is there a "crisis in expertise"? To answer that, we need to examine what expertise amounts to in a fragile world. Our guide must be the appropriate methodology, namely Case-Worker. What ought we expect from a scientist qua expert? After developing an answer, I will then situate it relative to existing work in the field.

To begin, Case-Worker rules out two popular conceptions of scientific expertise, or at least it shows those conceptions to be seriously incomplete.

11.3.1 Knowledge of theory

Does expertise consist in knowledge of theory? Theories are often seen as the goal of science and as the repository of scientific knowledge and achievement, and progress in science is seen as better theories. But as we see in other chapters, this view cannot survive the fact that when relations are fragile, theory plays only a toolbox role. Case-Worker demands supplementary investigations case by case to discover which theories apply, and scientific expertise must consist in part in expertise at these investigations, as well as in the local knowledge that they uncover. Theory alone is not enough.

Recent "debunking" studies make the point vivid. In Philip Tetlock's long-running forecasting tournaments of geopolitical and economic events, many credentialed experts score badly because they cling inflexibly to a single theory.[2] The forecasters who succeed, in contrast, as well as being numerate, are well informed and open-minded, and they update their forecasts continually by taking on board new information whenever relevant. These are Case-Worker virtues. They amount to skill at a version of contextual investigation. Expertise at forecasting goes far beyond mere knowledge of theories.

In another study, at the start of the COVID-19 pandemic, Cendri Hutcherson and colleagues asked scientists to predict the direction of change for a range of social and psychological phenomena.[3] The scientists' predictions turned out to be no better than those of a representative lay sample. It was not just prediction. Scientists also failed to outperform nonscientists in retrospective judgments of the same phenomena made six months into the pandemic. The scientists presumably have superior knowledge of theory. But without supplementary contextual investigations, this advantage did not lead to superior outcomes, either predictively or retrospectively.

Another recent study delivers the same message.[4] In various social domains, it was predicted how ideological preferences, political polarization, life satisfaction, sentiment on social media, and gender and racial bias, would change. Once again, theory-based forecasts performed badly—no better than lay forecasts—whereas forecasts made by those with more domain-specific knowledge performed better: "scientists were more accurate if they had scientific expertise in a prediction domain, were interdisciplinary, used simpler models and based predictions on prior data."[5]

Such studies show that theory alone is not enough. But they do not show—contrary to how they are sometimes reported—that science is fraudulent or in crisis. The key to seeing this is fragility. Subjects in these studies might be asked to estimate gross domestic product growth or the probability of an election result, for example. Many of the relations behind these target phenomena are fragile. Contextual knowledge is required. But the investigations required to gain that knowledge are exactly what, for any given topic, most participants in these studies have not done, so we should not expect accurate prediction or accurate retrospective analysis. Debunkers are themselves in the grip of unreasonable, fragility-ignoring expectations.

Analogously, imagine a historian who is an expert on the Aztecs. Should we put special weight on this historian's opinions about World War Two, medieval

[2] Tetlock (2005); Tetlock and Gardner (2015).
[3] Hutcherson et al. (2023).
[4] The Forecasting Collaborative (2023).
[5] The Forecasting Collaborative (2023).

chivalry, or the Industrial Revolution? Not unless they also have specialist knowledge about these further subjects. When relations are fragile, what is true for historians is true for scientists—expertise requires contextual knowledge over and above theory.

11.3.2 Knowledge of methods

Scientific objectivity, and thus scientific prestige, has often been defined in terms of method: once science is too large scale to rely on personal connections, best to trust instead impersonal procedures that all can follow.[6] Consider, for example, the detailed procedures formalized by the Evidence-Based Medicine and Evidence-Based Policy movements. Knowledge is defined as results obtained in approved ways, such as by randomized controlled trials or, better still, by meta-analyses of such trials. And expertise follows knowledge. Some anecdotes report how expertise is identified with mastery of methods, such as that funding committees spend disproportionate time on methods sections rather than on applications' substance.

But a methods conception of expertise fits badly with Case-Worker. If relations are fragile, emphasis shifts away from a-contextual methods and toward substantive, case-specific knowledge. Expertise still includes knowledge of methods, of course. But knowledge of methods alone, without substantive local knowledge to go with it, leaves an expert helpless, just as in the debunking studies. Better instead to be a "dirtbagger" to use Ashley Rubin's term: so long as you avoid obviously dubious practices, you may go easy on every detail of methods or formalisms and instead just opportunistically investigate when something interesting comes up, using whatever methods seem appropriate.[7]

11.4 A positive account: Case-Worker expertise

When relations are fragile, expertise means the ability to implement the Case-Worker strategy well. What does that consist in? In this section, I sketch an answer. Label this positive account "Case-Worker expertise."

Case-Worker favors knowledge of relevant theories and of methods by which to apply them, but it also favors knowledge of local circumstances. This

---

[6] Daston (1992).
[7] Rubin (2021).

last component is crucial. Without it, we cannot know which—if any—theories from our toolbox apply, and we cannot develop new theories effectively. Case-Worker expertise requires skill at contextual investigation in addition to knowledge of theory and methods, in a way that Stability-Theorist expertise does not.

Yet there is more to add. In addition to knowledge of theory, methods, and local circumstances, we need something further: skill at identifying what knowledge is needed, skill at getting it, and then skill at using it. This implies a certain kind of practical proficiency. It goes beyond what can be written down or formalized easily: the sensibility to react quickly when necessary, to know when and how to intervene effectively, to know what is and is not easily achievable, and to sense which variables and problems are fruitful. In addition to knowledge, this requires perhaps a good temperament and good intellectual habits, and an openness to Case-Worker rather than Stability-Theorist methods, much as Tetlock identified in his champion forecasters. A good case worker knows theory, *and* they know detailed facts of the case. They know a range of tools and rules of thumb, *and* they know how to draw from that range effectively.

In the insurance example, the case underwriter must integrate many different streams of evidence. This requires different skills from those of the flow underwriter, who manages an established statistical model. Similarly, the historian Ashworth could explain the World War One truces only by discovering, assessing, and integrating a huge amount of substantive knowledge of the case, not just by knowing familiar historical methods and social–psychological mechanisms.

Case-Worker expertise brings its own agenda. Consider meta-analyses: these are a staple of the Evidence-Based Medicine and Evidence-Based Policy movements, which consider the aggregation of results from randomized controlled trials to be the most authoritative of all evidence. But with fragility, aggregate figures are no longer reliable guides for action. We need to know instead what relations hold where and when. That means knowing support factors, derailers, and field guides to when they are likely to arise, and these in turn require contextual investigations and qualitative methods, not meta-analyses.[8]

Can there be Case-Worker expertise in fields that are only dubiously scientific, such as alien abductions or astrology? In one way, no. Expertise is a route to accurate predictions, interventions, and explanations, and alien abductions and astrology are dubious precisely because they lack these. (Arguably,

---

[8] Cartwright (2021); Pawson et al. (2005).

claims of alien appearances are deficient in part because they give up on the contextual detail characteristic of Case-Worker.[9]) But in another way, yes. We can still be expert much as an anthropologist or sociologist outsider can: in the second-order sense of knowing well these fields' internal debates and histories.

In one way, this positive account is disappointing. Expertise is contextual and therefore also narrow scope. Because no one can be an expert without fresh local spadework each time, we lose the comfort and balm of the all-knowing sage. But narrow-scope expertise that is genuine is better than wide-scope expertise that is illusory.

11.5 Other accounts of expertise

Case-Worker expertise may be clarified by comparing it to previous work. Begin with virtue epistemology. A strand in this literature emphasizes *under-standing* as a cognitive achievement distinct from knowledge.[10] Might expertise equate to understanding?

Jonathan Kvanvig distinguishes between "propositional" understanding of individual claims or theories and "objectual" understanding of wider entities such as a topic or subject matter.[11] Kvanvig relates this to coherence theories of justification: "understanding requires the grasping of explanatory and other coherence-making relationships in a large and comprehensive body of information ... What is distinctive about understanding [compared to knowledge] has to do with the way in which an individual combines pieces of information into a unified body."[12]

Objectual understanding has since been developed by Catherine Elgin, who sees it as a central goal of science.[13] In Elgin's words, objectual understanding of chemistry, for example, means:

[understanding] why various chemical reactions occur, why various chemicals bond, and so forth. But it [also means that one] . . . appreciates how a variety of epistemic commitments hang together in a mutually supportive network. These commitments are not just statements of fact; they include methods for assessing whether particular facts hold, whether they are

---

[9] Turner and Turner (2021).
[10] Hannon (2021) surveys recent work.
[11] Kvanvig (2003).
[12] Kvanvig (2003, 192, 197).
[13] Elgin (2017, 2018).

relevant, and whether they support each other, as well as orientations toward the phenomena, and standards of acceptability that determine whether the system as a whole is worthy of reflective endorsement.[14]

Objectual understanding is clearly similar in spirit to Case-Worker expertise. The two share a holistic emphasis on how things hang together. And Kvanvig's and Elgin's distinction between understanding and knowledge has some overlap with my distinction between expertise and knowledge of theory.

Elgin is motivated by the key role in science of idealized models: roughly, such models are literally false, so scientific achievement cannot be based on knowledge of them because knowledge implies facticity, but basing scientific achievement on understanding avoids this problem. The arguments in this book for Case-Worker expertise, in contrast, are not motivated by idealized models. Fragility therefore offers a new rationale for objectual understanding.

Another forebear from epistemology is the notion of "understanding-why." In Alison Hills's formulation, this amounts to a kind of "cognitive control" because, roughly, it enables a deeper and wider appreciation of how claims hang together than mere knowledge does.[15] As Hills puts it, understanding-why "is useful . . . when you need to tackle a new question because it guarantees that you have the know-how to do this successfully."[16]

Understanding-why and objectual understanding offer some of the benefits of causal knowledge, which, according to manipulationist views of causation, boils down to knowledge of the network of causal dependency relations in a target.[17] Just as mechanisms do, such networks tell us the underpinnings of surface-level explanatory claims. Causal knowledge does capture some of Case-Worker expertise: contextual investigations often aim to discover a target's network of causal dependency relations. Case-Worker expertise, though, is not just that. It is also something broader—skill at judging which contextual investigations to undertake in the first place and then skill at doing them. This is part of what distinguishes Case-Worker from Stability-Theorist. Expertise is not just knowledge of networks of causal dependency relations; it is also the skill and wisdom that must be deployed to gain such knowledge. This is what marks out Tetlock's superforecasters, too.

Understanding is also a rising topic in philosophy of science.[18] Recent work addresses the relation between understanding and explanation, and it sees

---

[14] Elgin (2018, 333).
[15] Hills (2016).
[16] Hills (2016, 678).
[17] Woodward (2003); Gopnik and Schulz (2007).
[18] De Regt (2017); Khalifa (2017); Potochnik (2017); Humphreys (2000).

understanding as a central—and distinct—goal of science. Henk de Regt develops the view that, roughly, to understand a phenomenon is to have an explanation of it that is based on an intelligible theory. Intelligibility, in turn, is a cluster of qualities that facilitate a theory's use. De Regt's notion of understanding clearly overlaps with the notions of objectual understanding and understanding-why from epistemology. To capture Case-Worker expertise fully, I would want to add extra practical dimensions, as well as, in de Regt's case, to tie the notion less closely to theory, given the heightened role in Case-Worker for extra-theoretical components.

Hasok Chang develops in detail a pragmatist account of expertise that overlaps with Case-Worker expertise.[19] He argues that the essence of scientific achievement is the skill to do things, rather than just to know propositions or theories, and so a central part of expertise must be proficiency at epistemic activities and practices: "knowledge-as-ability" in Chang's phrase. This view is close to my own.

As remarked earlier, Nancy Cartwright and coauthors' notion of the "tangle" of science also gets at a lot of Case-Worker expertise.[20] Cartwright writes: "[theories] are a boon. We do not need to recast them as propositions, we do not need to take them literally, we do not need to try to 'derive' our conclusions from them. We do need to learn how to use them. We do need, as a community, to fill our toolbox with a big variety of tools and to learn how to use them together to craft reliable finished products."[21] Case-Worker expertise is knowing how to do this.

Might Case-Worker expertise correspond to knowing-how and knowledge to knowing-that? I do not think so. Case-Worker expertise includes plenty of propositional knowing-that—the Case-Worker expert generally does not solve problems just by unarticulated instinct or learning. In this way, expertise diverges from some colloquial usages of "expert," as when we say that "Federer is an expert tennis player." It remains true that knowing-how can form *part* of expertise, as when the know-how of a laboratory technician forms part of a *team's* expertise, or as when, in earlier times, much of a scientist's expertise might have consisted in their know-how as an experimentalist or instrument-maker.

Something like Case-Worker expertise has often appealed to practitioners—not surprisingly. Here are two examples. First, Nobel Prize–winning economist Esther Duflo urges economists to think of themselves as like plumbers

[19] Chang (2012, 2022); see also Westerblad (2023).
[20] Cartwright et al. (2022).
[21] Cartwright (2020a, 319).

who, to get things done, need not just knowledge of theory but also knowledge of contextual, practical details.[22] Other economists urge comparison to an engineer or to an experienced craftsman.[23] Don Ross, too, champions an "engineering" role for economics; unlike me, he sees this role as largely unaffected by the methodology of economic theory development.[24] Duflo herself is famous for pioneering the use of randomized field trials to test development policies.[25]

All of this is laudable. It encourages a conception of expertise akin to Case-Worker expertise; it emphasizes implementing rather than theorizing. Ideally, it would go further. Why restrict Duflo's economist-plumber to a narrow range of methods (just randomized trials for causal inference), a narrow range of source theories (just models taken from orthodox economics), and a narrow range of implementation tools (mainly just incentive effects)? A richer expertise would embrace pluralism about all of these.

The second example is Donald Schön's highly influential book, *The Reflective Practitioner*.[26] He examines the professions of engineering, architecture, management, town planning, and psychotherapy, with briefer coverage too of nursing, teaching, and others. Schön argues that practitioners in these fields face "situations of uncertainty, instability, uniqueness and conflict."[27] How best to respond? By selecting intelligently from a toolbox of theories and methods, case by case. In our terminology, practitioners face fragility and should respond by following Case-Worker. Schön's main foil is "technical rationality," by which he means theories that are both developed and applied abstractly rather than empirically—akin to Stability-Theorist. These theories are too abstract to be helpful, according to him: cognitive psychology theories are unhelpful to teachers, policy science theories are unhelpful to policy administrators, and management science theories are unhelpful to managers. This leaves practitioners with a dilemma—in Schön's phrase, "rigor or relevance?"

Strikingly, the solution, Schön argues, is not only the Case-Worker strategy for theory application but also the Case-Worker strategy for theory *development*, just as urged in Chapter 4. Field work and case studies are crucial to that. "The development of action science cannot be achieved by researchers who keep themselves removed from contexts of action ... Field work, consultation,

[22] Duflo (2017).
[23] Roth (2002); Banerjee (2002).
[24] Ross (2014).
[25] Banerjee and Duflo (2012).
[26] Schön (1992)—in its various editions, cited over 87,000 times, according to Google Scholar.
[27] Schön (1992, 308).

and continuing education, often considered as second-class activities or as necessary evils, will rise to first-class status as vehicles of research, the main business of the university."[28] Schön favors "repertoire-building research"—in our terminology, expanding the toolbox of theories.

Expertise in the professions, then, on Schön's account, consists in skill at developing and bringing together methods and theories in a contextually sensitive way. This is Case-Worker expertise. Similar claims are found in other reflective works by practitioners, such as Kathryn Montgomery's *How Doctors Think*.[29] And the novelist Leo Tolstoy, too, endorsed something like Case-Worker expertise. At one time, he set up many schools on his estate, and about teaching children to read, he wrote:

> The best teacher will be he who has at his tongue's end the explanation of what it is that is bothering the pupil. These explanations give the teacher knowledge of the greatest possible number of methods, the ability to invent new methods and, above all, not a blind adherence to one method but the conviction that all methods are one-sided, and that the best method would be the one which would answer best to all the possible difficulties incurred by a pupil, that is, not a method but an art and talent.[30]

Case-Worker expertise has a *social* element. Just as some individuals are good at producing accurate predictions, explanations, and interventions when faced with fragility, so too are some communities and social arrangements. We need good epistemic environments. Alvin Goldman defines an expert to be someone who can solve problems or execute tasks for someone else who cannot.[31] Other work in social epistemology gives more detail. According to Helen Longino, for science to generate knowledge effectively, it needs cognitive and social diversity, effective critical interaction in and between epistemic communities, reliability of reasoning, and acceptability of values.[32] How should different epistemic communities combine? Longino advocates "interactive objectivity" to bring together narrow areas of expertise productively. Peter Galison uses the metaphor of a "trading zone" to explain how physicists from different paradigms collaborate with each other and with engineers to develop high-energy physics particle detectors.[33] Harry Collins develops the

[28] Schön (1992, 320–4).
[29] Montgomery (2006).
[30] Tolstoy (1967/1862), quoted in Schön (1992, 66).
[31] Goldman (2018).
[32] Longino (1990, 2002).
[33] Galison (1997).

idea of an "interactional expert"—someone trained in different disciplines and areas of expertise, who is able to exchange information and communicate with different experts.[34] In all these cases, the conception of expertise is broad, and well suited to Case-Worker.

Many previous discussions of expertise focus on its political aspects, especially on how assertions of expertise serve to prioritize the agendas of those deemed to be the experts. Donald Schön, for example, emphasizes that frames imposed by experts assert certain values.[35] Many critiques of economists make a similar point.

How does fragility bear on these political aspects? The main connection is that when relations are fragile, theories are often not epistemically decisive. No successful experiment or working artifact settles matters and, compared to when relations are stable, it is more controversial what the correct theory is, or what the best explanation or intervention is. The verdict of science is less clear. This is a double-edged sword. On one hand, it offers the prospect of a more humane science, no longer rigidly beholden to dominant theories. On the other hand, it eases entry for dubious influences and for bad-faith manipulation of "what science says." Many ways in which sexism is alleged by feminists to enter science, for example, are via underdetermination of theory, and fragility makes such underdetermination more salient.[36] Case-Worker favors the local investigations typical of a historian. It is no accident that a historian's sensibility is more attuned to political influences than is a theorist's.

Conversely, appeals to stability, too, can be abused. With stable relations comes epistemic authority, because those relations can be relied on to apply—which is fine unless, in fact, those relations are not stable, or unless although stable they are outweighed by other relations, or unless although stable they depend on many other fragile relations to be implemented effectively. James C. Scott's well-known criticisms of government allege this pattern. Roughly, he objects to authoritarian implementation of (allegedly) stable interventions that lead to disaster because they ignore local knowledge.[37]

What of political change? Fragility is especially common in social domains. Case-Worker expertise suggests an enlarged role for activists, as opposed to social scientists, because academic training is less essential for accumulating local knowledge. There is more of a gap for "amateurs." If academics do not fill

---

[34] Collins and Evans (2007).
[35] Schön (1992).
[36] Longino (1990); Okruhlik (1994).
[37] Scott (1999).

this gap, then others will. In good cases, that might mean think tanks or well-informed journalists, in bad cases, less savory sources.

11.6   Who is the audience for the book?

According to this book, science sometimes goes astray by pursuing Stability-Theorist even when relations are fragile. Why is this mistake made? That is a very important question. Any serious answer to it requires serious investigation, and this answer will likely vary case by case—perhaps different mixtures of the prestige of stability solutions, funding bodies and teaching practices that endorse this prestige, and much else besides.

How might we institutionalize Case-Worker expertise? Given the norms that currently govern many sciences, it will be a difficult job to reorient from theory and methods toward a more fragility-suitable conception—for example, toward norms of publication that favor investigation of support factors and derailers, rather than collection of average treatment effects from randomized controlled trials or meta-analyses of them. Reform requires detailed study of the causes of error—in other words, it requires an answer to the same question as in the previous paragraph.

That answer also bears on what lessons the book carries for different audiences: who has a role to play in any reform? I think all of the following do. Philosophers of science—do not peddle myths that support bad science—recognize and promote good science. Science funders and planners—fund science that is done right. Practitioners—decide on the right methods to follow or to support. Science educators and trainers—put the right kind of science on curricula and emphasize the skills and sensitivities conducive to the right kind of expertise.

I have not investigated here why science sometimes goes astray in the way I say it does. Before doing that, we must first establish the prior claim that science *does* sometimes go astray in the way I say it does. That is a goal of this book. Even that is job enough.

Bibliography

Achinstein, Peter. 1983. *The Nature of Explanation*. Oxford: Oxford University Press.

Ackroyd, Stephen. 2009. "Research Designs for Realist Research." In *The Sage Handbook of Organizational Research Methods*, edited by David Buchanan and Alan Bryman, 532–48. London: Sage.

Alexandrova, Anna. 2008. "Making Models Count." *Philosophy of Science* 75, 383–404.

Alexandrova, Anna, and Robert Northcott. 2009. "Progress in Economics: Lessons from the Spectrum Auctions." In *The Oxford Handbook of Philosophy of Economics*, edited by Harold Kincaid and Don Ross, 306–37. Oxford University Press.

Alexandrova, Anna, Robert Northcott, and Jack Wright. 2021. "Back to the Big Picture." *Journal of Economic Methodology* 28 (1), 54–9.

Anderson, Chris. 2008. "The End of Theory: The Data Deluge Makes the Scientific Method Obsolete." *Wired Magazine*, June 23.

Angner, Erik. 2019. "We're All Behavioral Economists Now." *Journal of Economic Methodology* 26 (3), 195–207.

Angrist, Joshua, Pierre Azoulay, Glenn Ellison, Ryan Hill, and Susan Feng Lu. 2017. "Empirical Research Evolves: Fields and Styles." *American Economic Review* 107 (5), 293–7.

Angrist, Joshua, and Jörn-Steffen Pischke. 2009. *Mostly Harmless Econometrics: An Empiricist's Companion*. Princeton: Princeton University Press.

Angrist, Joshua, and Jörn-Steffen Pischke. 2010. "The Credibility Revolution in Empirical Economics: How Better Research Design Is Taking the Con out of Econometrics." *Journal of Economic Perspectives* 24, 3–30.

Ankeny, Rachel. 2012. "Using Cases to Establish Novel Diagnoses: Creating Generic Facts by Making Particular Facts Travel Together." In *How Well Do Facts Travel?*, edited by Peter Howlett and Mary Morgan, 252–72 . Cambridge: Cambridge University Press.

Asay, Jamin. 2019. "Going Local: A Defense of Methodological Localism about Scientific Realism." *Synthese* 196, 587–609.

Ashworth, Tony. 1980. *Trench Warfare 1914–1918*. MacMillan.

Axelrod, Robert. 1984. *The Evolution of Cooperation*. Penguin.

Aydinonat, N. Emrah. 2008. *The Invisible Hand in Economics: How Economists Explain Unintended Social Consequences*. London: Routledge.

Banerjee, Abhijit. 2002. "The Uses of Economic Theory: Against a Purely Positive Interpretation of Theoretical Results." Unpublished manuscript, available at SSRN: https://ssrn.com/abstract = 315942 or http://dx.doi.org/10.2139/ssrn.315942

Banerjee, Abhijit, and Esther Duflo. 2012. *Poor Economics: A Radical Rethinking of the Way to Fight Global Poverty*. New York: PublicAffairs.

Barrotta, Pierluigi, and Eleonora Montuschi. 2018. "The Dam Project: Who are the Experts? A Philosophical Lesson from the Vajont Disaster." In *Science and Democracy. Controversies and Conflicts*, edited by Pierluigi Barrotta and Giovanni Scarafile, 17–33. Amsterdam: Benjamins.

BBC. 2020. "Coronavirus: 'Sombre day' as UK Deaths Hit 10,000." BBC News online, April 12.

Bechtel, William, and Robert Richardson. 2010. *Discovering Complexity: Decomposition and Localization as Strategies in Scientific Research.* MIT Press.

Beckage, Brian, Stuart A. Kauffman, Lou Gross, Asim Zia, and Christopher J. Koliba. 2013. "More Complex Complexity: Exploring the Nature of Computational Irreducibility Across Physical, Biological, and Human Social Systems." In *Irreducibility and Computational Equivalence*, edited by Hector Zenil, 79–88. Berlin Heidelberg: Springer-Verlag.

Becker, Gary. 1976. *The Economic Approach to Human Behavior.* Chicago: Chicago University Press.

Becker, Howard. 2014. *What about Mozart? What about Murder? Reasoning from Cases.* Chicago: University of Chicago Press.

Beebe, James, and Finnur Dellsén. 2020. "Scientific Realism in the Wild: An Empirical Study of Seven Sciences and History and Philosophy of Science." *Philosophy of Science* 87 (2), 336–64.

Ben-Menahem, Yemima. 1997. "Historical Contingency." *Ratio* 10, 99–107.

Ben-Menahem, Yemima. 2009. "Historical Necessity and Contingency." In *A Companion to the Philosophy of History and Historiography*, edited by Aviezer Tucker, 120–30. Chichester, England: Wiley-Blackwell.

Bengoetxea, Juan Bautista, and Oliver Todt. 2021. "Decision Making in the Nutrition Sciences." *Manuscrito* 44 (3), 42–69.

Berk, Richard A., and David A. Freedman. 2003. "Statistical Assumptions as Empirical Commitments." In *Law, Punishment, and Social Control: Essays in Honor of Sheldon Messinger*, edited by Stanley Cohen, 235–54. New York: Aldine.

Betz, Gregor. 2006. *Prediction or Prophecy?* Wiesbaden: Deutscher Universitaets Verlag.

Bhaskar, Roy. 1975. *A Realist Theory of Science.* Leeds: Leeds Books.

Biddle, Jeff E., and Daniel S. Hamermesh. 2016. "Theory and Measurement: Emergence, Consolidation and Erosion of a Consensus." NBER Working Paper 22253.

Billah, Arif, Mamun Miah, and Nuruzzaman Khan. 2020. "Reproductive Number of Coronavirus: A Systematic Review and Meta-Analysis Based on Global Level Evidence." *PLoS ONE* 15 (11), e0242128. November 11.

Binmore, Ken. 1994. *Game Theory and the Social Contract, Vol. 1: Playing Fair.* Cambridge, MA: MIT Press.

Binmore, Ken. 1998. *Game Theory and the Social Contract, Vol. 2: Just Playing.* Cambridge, MA: MIT Press.

Bird, Alexander. 2007. "What Is Scientific Progress?" *Nous* 41 (1), 64–89.

Bohrnstedt, George, and Brian Stecher (eds). 2002. *What We have Learned about Class Size Reduction in California.* Sacramento: California Department of Education.

Boon, Mieke. 2020. "How Scientists are Brought Back into Science: The Error of Empiricism." In *A Critical Reflection on Automated Science: Will Science Remain Human? (Vol. 1)*, edited by Marta Bertolaso and Fabio Sterpetti, 43–65. Springer Cham.

Boumans, Marcel. 2015. *Science Outside the Laboratory: Measurement in Field Science and Economics.* New York: Oxford University Press.

Brady, Henry, and David Collier (eds). 2010. *Rethinking Social Inquiry: Diverse Tools, Shared Standards.* Lanham: Rowman and Littlefield.

Brante, Thomas. 2001. "Consequences of Realism for Sociological Theory-Building." *Journal for the Theory of Social Behaviour* 31, 167–94.

Brazeau, N., R. Verity, S. Jenks, H. Fu, C. Whittaker, P. Winskill, I. Dorigatti, P. Walker, S. Riley, R. Schenkenberg, H. Hoelgebaum, T. Mellan, S. Mishra, J. Unwin, O. Watson, Z. Cucunuba, M. Baguelin, L. Whittles, S. Bhatt, A. Ghani, N. Ferguson, and L. Okell. 2020.

"Report 34—COVID-19 Infection Fatality Ratio Estimates from Seroprevalence." *MRC Centre for Global Infectious Disease Analysis*, October 29.

Broadbent, Alex. 2013. *Philosophy of Epidemiology*. Palgrave MacMillan.

Broadbent, Alex, and Benjamin Smart. 2020. "Why a One-Size-Fits-All Approach to COVID-19 Could Have Lethal Consequences." *The Conversation* website, March 23.

Bryman, Alan. 2016. *Social Research Methods*. Oxford: Oxford University Press.

Calude, Cristian, and Giuseppe Longo. 2017. "The Deluge of Spurious Correlations in Big Data." *Foundations of Science* 22 (3), 595–612.

Card, David, and Alan B. Krueger. 1994. "Minimum Wages and Employment: A Case Study of the Fast Food Industry in New Jersey and Pennsylvania." *American Economic Review* 84, 772–93.

Cartwright, Nancy. 1983. *How the Laws of Physics Lie*. Oxford University Press.

Cartwright, Nancy. 1989. *Nature's Capacities and Their Measurement*. Oxford: Oxford University Press.

Cartwright, Nancy. 1999. *The Dappled World: A Study of the Boundaries of Science*. Cambridge: Cambridge University Press.

Cartwright, Nancy. 2012. "Will This Policy Work for You? Predicting Effectiveness Better: how Philosophy Helps." *Philosophy of Science* 79, 973–89.

Cartwright, Nancy. 2013. "Evidence, Argument and Prediction." In *EPSA11 Perspectives and Foundational Problems in Philosophy of Science, the European Philosophy of Science Association Proceedings 2*, edited by Vassilios Karakostas and Dennis Dieks, 3–17. Basel: Springer.

Cartwright, Nancy. 2019. *Nature, the Artful Modeler*. Chicago: Open Court.

Cartwright, Nancy. 2020a. "Middle-Range Theory: Without It What Could Anyone Do?" *Theoria* 35 (3), 269–323.

Cartwright, Nancy. 2020b. "Why Trust Science? Reliability, Particularity, and the Tangle of Science." *Proceedings of the Aristotelian Society* 120 (3), 237–52.

Cartwright, Nancy. 2021. "Rigour versus the Need for Evidential Diversity." *Synthese* 199, 13095–119.

Cartwright, Nancy, Lucy Charlton, Matt Juden, Tamlyn Munslow, and Richard Beadon Williams. 2020. "Making Predictions of Programme Success More Reliable." *CEDIL Methods Working Paper 1*. Oxford: Centre of Excellence for Development Impact and Learning.

Cartwright, Nancy, and Jeremy Hardie. 2012. *Evidence-Based Policy: A Practical Guide to Doing It Better*. Oxford: Oxford University Press.

Cartwright, Nancy, Jeremy Hardie, Eleonora Montuschi, Matthew Soleiman, and Ann C. Thresher. 2022. *The Tangle of Science*. Oxford: Oxford University Press.

Cartwright, Nancy, John Pemberton, and Sarah Wieten. 2018. "Mechanisms, Ceteris Paribus Laws and Covering-Law Explanation." Centre for Philosophy of Natural and Social Science Working Paper. London: London School of Economics.

Cartwright, Nancy, Towfic Shomar, and Mauricio Suarez. 1995. "The Tool Box of Science: Tools for the Building of Models with a Superconductivity Example." *Poznan Studies in the Philosophy of the Sciences and the Humanities* 44, 137–149.

Chakravartty, Anjan. 2017. "Scientific Realism." In *The Stanford Encyclopedia of Philosophy* (Summer 2017 Edition), edited by Edward N. Zalta. https://plato.stanford.edu/archives/sum2017/entries/scientific-realism/

Chang, Hasok. 2004. *Inventing Temperature*. New York: Oxford University Press.

Chang, Hasok. 2012. *Is Water H2O? Evidence, Realism and Pluralism*. Dordrecht: Springer.

Chang, Hasok. 2022. *Realism for Realistic People: A New Pragmatist Philosophy of Science*. Cambridge: Cambridge University Press.

Chapman, Robert, and Alison Wylie. 2016. *Evidential Reasoning in Archaeology*. London: Bloomsbury.

Chen, Jeff. 2021. "Thinking Ahead: Why the Mathematics of Stability Matters for Policy." Bennett Institute for Public Policy Cambridge blog, March 31. Thinking ahead: Why the mathematics of stability matters for policy—Bennett Institute for Public Policy (cam. ac.uk)

Colander, David, and Harry Landreth. 2004. "Pluralism, Formalism and American Economics." Middlebury College Economics Discussion Paper 04–09.

Collier, David. 2011. "Understanding Process Tracing." *PS: Political Science and Politics* 44 (4), 823–30.

Collins, Harry, and Robert Evans. 2007. *Rethinking Expertise*. Chicago and London: The University of Chicago Press.

Conn, David, Felicity Lawrence, Paul Lewis, Severin Carrell, David Pegg, Harry Davies, and Rob Evans. 2020. "Revealed: The Inside Story of the UK's Covid-19 Crisis." *The Guardian*, April 29. https://www.theguardian.com/world/2020/apr/29/revealed-the-ins ide-story-of-uk-covid-19-coronavirus-crisis

Cooper, Rachel. 2004. "Why Hacking Is Wrong about Human Kinds." *British Journal for the Philosophy of Science* 55, 73–85.

da Costa, Newton, and Steven French. 2003. *Science and Partial Truth: A Unitary Approach to Models and Reasoning in Science*. New York: Oxford University Press.

Costello, Anthony. 2020. "The Government's Secret Science Group has a Shocking Lack of Expertise." *The Guardian*, April 27. The government's secret science group has a shocking lack of expertise | Coronavirus | The Guardian

Craver, Carl. 2007. *Explaining the Brain: Mechanisms and the Mosaic Unity of Neuroscience*. New York: Oxford University Press.

Craver, Carl and James Tabery, "Mechanisms in Science." In *The Stanford Encyclopedia of Philosophy (Summer 2019 Edition)*, edited by Edward N. Zalta.https://plato.stanford. edu/archives/sum2019/entries/science-mechanisms/.

Currie, Adrian. 2014. "Narratives, Mechanisms and Progress in Historical Science." *Synthese* 191 (6), 1163–83.

Currie, Adrian. 2019a. *Scientific Knowledge and the Deep Past: History Matters*. Elements in the Philosophy of Science. Cambridge: Cambridge University Press.

Currie, Adrian. 2019b. "Mass Extinctions as Major Transitions." *Biology and Philosophy* 34 (2):29.

Currie, Adrian, and Kirsten Walsh. 2018. "Newton on Islandworld: Ontic-Driven Explanations of Scientific Method." *Perspectives on Science* 26 (1), 119–56.

Daston, Lorraine. 1992. "Objectivity and the Escape from Perspective." *Social Studies of Science* 22 (4), 597–618.

Davidson, Donald. 1963. "Actions, Reasons and Causes." *Journal of Philosophy* 60, 685–700.

Deaton, Angus, and Nancy Cartwright. 2018. "Understanding and Misunderstanding Randomized Controlled Trials." *Social Science and Medicine* 210, 2–21.

de Regt, Henk. 2017. *Understanding Scientific Understanding*. Oxford University Press.

Diener, Ed, Rober Northcott, Michael J. Zyphur, and Stephen G. West. 2022. "Beyond Experiments." *Perspectives on Psychological Science* 17 (4), 1101–19.

Dowding, Keith. 2016. *The Philosophy and Methods of Political Science*. Basingstoke: Palgrave Macmillan.

Dowding, Keith, and Charles Miller. 2019. "On Prediction in Political Science." *European Journal of Political Research* 58 (3), 1001–18.

Dresow, Max. 2021. "Explaining the Apocalypse: The End-Permian Mass Extinction and the Dynamics of Explanation in Geohistory." *Synthese* 199, 10441–74.

Dretske, Fred. 1972. "Contrastive Statements." *Philosophical Review* 81 (4), 411–37.

Duflo, Esther. 2017. "The Economist as Plumber." *American Economic Review* 107 (5), 1–26.

Dupré, John. 1993. *The Disorder of Things: Metaphysical Foundations of the Disunity of Science.* Harvard University Press.

Dupré, John. 2012. *Processes of Life: Essays in Philosophy of Biology.* Oxford: Oxford University Press.

Dupré, John. 2013. "Living Causes." *Proceedings of the Aristotelian Society* 87 (1), 19–37.

Dupré, John. 2021. *The Metaphysics of Biology.* Elements in the Philosophy of Biology. Cambridge: Cambridge University Press.

Dupré, John, and Stephan Guttinger. 2016. "Viruses as Living Processes." *Studies in History and Philosophy of Science Part C: Studies in History and Philosophy of Biological and Biomedical Sciences* 59, 109–16.

Durkheim, Emile. 1897/1951. *Suicide: A Study in Sociology.* Translated by John Spaulding and George Simpson. Glencoe: Free Press.

Edeling, W., A. Hamid, R. Sinclair, D. Suleimenova, K. Gopalakrishnan, B. Bosak, D. Groen, I. Mahmood, D. Crommelin, and P. Coveney. 2020. "Model Uncertainty and Decision Making: Predicting the Impact of COVID-19 Using the CovidSim Epidemiological Code." October 7. DOI: 10.21203/rs.3.rs-82122/v1. [PDF] Model uncertainty and decision making: Predicting the Impact of COVID-19 Using the CovidSim Epidemiological Code | Semantic Scholar

Elgin, Catherine. 2017. *True Enough.* Cambridge: MIT Press.

Elgin, Catherine. 2018. "Epistemic Virtues in Understanding." In *Routledge Handbook of Virtue Epistemology*, edited by Heather D. Battaly, 330–9. London: Routledge.

Elliott-Graves, Alkistis. 2016. "The Problem of Prediction in Invasion Biology." *Biology and Philosophy* 31, 373–93.

Elliott-Graves, Alkistis. 2018. "Generality and Causal Interdependence in Ecology." *Philosophy of Science* 85, 1102–14.

Elliott-Graves, Alkistis. 2019. "The Future of Predictive Ecology." *Philosophical Topics* 47, 65–82.

Elster, Jon. 1988. "The Nature and Scope of Rational-Choice Explanation." In *Science in Reflection*, edited by Edna Ullmann-Margalit, 51–65. Springer Netherlands.

Elster, Jon. 1989. *Nuts and Bolts for the Social Sciences.* Cambridge: Cambridge University Press.

Elster, Jon. 2015. *Explaining Social Behavior. More Nuts and Bolts for the Social Sciences.* Cambridge, UK: Cambridge University Press.

Epstein, Brian. 2014. "Why Macroeconomics Does Not Supervene on Microeconomics." *Journal of Economic Methodology* 21, 3–18.

Epstein, Brian. 2015. *The Ant Trap: Rebuilding the Foundations of the Social Sciences.* Oxford: Oxford University Press.

Espeland, Wendy, and Michael Sauder. 2016. *Engines of Anxiety: Academic Rankings, Reputation, and Accountability.* New York: Russell Sage Foundation.

Farmer, J. Doyne. 2013. "Hypotheses Non Fingo: Problems with the Scientific Method in Economics." *Journal of Economic Methodology* 20 (4), 377–85.

Ferguson, N., D. Laydon, G. Nedjati-Gilani, N. Imai, K. Ainslie, M. Baguelin, S. Bhatia, A. Boonyasiri, Z. Cucunubá, G. Cuomo-Dannenburg, A. Dighe, I. Dorigatti, H. Fu, K. Gaythorpe, W. Green, A. Hamlet, W. Hinsley, L. Okell, S. van Elsland, H. Thompson, R. Verity, E. Volz, H. Wang, Y. Wang, P. Walker, C. Walters, P. Winskill, C. Whittaker, C. Donnelly, S. Riley, A. Ghani. 2020. "Report 9—Impact of Non-pharmaceutical Interventions (NPIs) to Reduce COVID-19 Mortality and Healthcare Demand." *MRC Centre for Global Infectious Disease Analysis*, March 16.

Forber, Patrick. 2010. "Confirmation and Explaining How Possible." *Studies in History and Philosophy of Science Part C* 41, 32–40.

Ford, Jonathan. 2020. "The Battle at the Heart of British Science Over Coronavirus." *Financial Times*, April 15. https://www.ft.com/content/1e390ac6-7e2c-11ea-8fdb-7ec06 edeef84?desktop = true&segmentId = 7c8f09b9-9b61-4fbb-9430-9208a9e233c8

The Forecasting Collaborative. 2023. "Insights into the Accuracy of Social Scientists' Forecasts of Societal Change." *Nature Human Behaviour,* February 9. (130 co-authors)

Foster, Ian, Rayid Ghani, Ron S. Jarmin, Frauke Kreuter, and Julia Lane. 2017. *Big Data and Social Science.* Boca Raton: CRC Press.

Franklin, Allan, and Slobodan Perovic. 2021. "Experiment in Physics." *The Stanford Encyclopedia of Philosophy* Summer 2021 Edition, edited by Edward N. Zalta.

French, Steven. 2020. *There are No Such Things as Theories.* Oxford: Oxford University Press.

Friedman, Michael. 1974. "Explanation and Scientific Understanding." *Journal of Philosophy* 71 (1), 5–19.

Friedman, Milton. 1953. "The Methodology of Positive Economics." In *Essays in Positive Economics*, by Milton Friedman, 3–43. Chicago: University of Chicago Press.

Frigg, Roman, and Stephan Hartmann. 2020. "Models in Science." In *The Stanford Encyclopedia of Philosophy (Spring 2020 Edition)*, edited by Edward N. Zalta. https://plato.stanford.edu/archives/spr2020/entries/models-science/.

Fuller, Jonathan. 2021. "What Are COVID-19 Models Modeling (Philosophically Speaking)?" *History and Philosophy of the Life Sciences* 43 (2), 1–5.

Galison, Peter. 1997. *Image and Logic. A Material Culture of Microphysics.* Chicago: The University of Chicago Press.

Garfinkel, Alan. 1981. *Forms of Explanation.* New Haven: Yale University Press.

Geertz, Clifford. 1973. "Thick Description: Toward an Interpretive Theory of Culture." In *The Interpretation of Cultures*, by Clifford Geertz, 3–30. New York: Basic Books.

Gelman, Andrew. 2008. "Methodology as Ideology: Some comments on Robert Axelrod's *The Evolution of Co-operation*." *QA-Rivista dell'Associazione Rossi-Doria*, 167–76.

Gennetian, Lisa A., Lisa Sanbonmatsu, Lawrence F. Katz, Jeffrey R. Kling, Matthew Sciandra, Jens Ludwig, Greg Duncan, and Ronald C. Kessler. 2012. "The Long-term Effects of Moving to Opportunity on Youth Outcomes." *Cityscape* 14 (2), 137–67.

Ghomi, Hamed Tabatabaei. 2022. "Setting the Demons Loose: Computational Irreducibility Does Not Guarantee Unpredictability or Emergence." *Philosophy of Science* 89 (4), 761–83.

Gintis, Herbert. 2007. "A Framework for the Unification of the Behavioral Sciences." *Behavioral and Brain Sciences* 30 (1), 1–61.

Gintis, Herbert. 2016. *Individuality and Entanglement: The Moral and Material Bases of Social Life.* Princeton: Princeton University Press.

Glennan, Stuart. 2017. *The New Mechanical Philosophy.* New York: Oxford University Press.

Goertz, Gary. 2012. "Descriptive-Causal Generalizations: "Empirical Laws" in the Social Sciences." In *The Oxford Handbook of Philosophy of Social Science*, edited by Harold Kincaid, 85–108. Oxford University Press.

Goertz, Gary, and James Mahoney. 2012. *A Tale of Two Cultures: Qualitative and Quantitative Research in the Social Sciences.* Princeton: Princeton University Press.

Goldman, Alvin. 2018. "Expertise." *Topoi* 37 (1), 3–10.

Gopnik, Alison, and Laura Schulz (eds). 2007. *Causal Learning: Psychology, Philosophy, and Computation.* New York: Oxford University Press.

Gowa, Joanne. 1986. "Anarchy, Egoism, and Third Images: The Evolution of Cooperation and International Relations." *International Organization* 40 (1), 167–86.

Greenhalgh, Trisha. 2020. "Will COVID-19 Be Evidence-Based Medicine's Nemesis?" *plos. org*, June 30. https://doi.org/10.1371/journal.pmed.1003266

Greenhalgh, Trisha, and Chrysanthi Papoutsi. 2018. "Studying Complexity in Health Services Research: Desperately Seeking an Overdue Paradigm Shift." *BMC Medicine* 16 (1). DOI:10.1186/s12916-018-1089-4

Grépin, Karen Ann, Tsi-Lok Ho, Zhihan Liu, Summer Marion, Julianne Piper, Catherine Z. Worsnop, and Kelley Lee. 2021. "Evidence of the Effectiveness of Travel-Related Measures During the Early Phase of the COVID-19 Pandemic: A Rapid Systematic Review." *BMJ Global Health* 6, e004537.

Grey, Stephen, and Andrew MacAskill. 2020. "Special Report: Johnson Listened to His Scientists About Coronavirus—But They Were Slow to Sound the Alarm." *Reuters*, April 7. Special Report: Johnson listened to his scientists about coronavirus - but they were slow to sound the alarm | Reuters

Grüne-Yanoff, Till. 2009. "Learning from Minimal Economic Models." *Erkenntnis* 70, 81–99.

Grüne-Yanoff, Till. 2016. "Why Behavioral Policy Needs Mechanistic Evidence." *Economics and Philosophy* 32.3, 463–83.

Guala, Francesco. 2005. *Methodology of Experimental Economics*. Cambridge: Cambridge University Press.

Hacking, Ian. 1983. *Representing and Intervening: Introductory Topics in the Philosophy of Natural Science*. Cambridge: Cambridge University Press.

Hacking, Ian. 1995. "The Looping Effects of Human Kinds." In *Causal Cognition: A Multidisciplinary Debate*, edited by Dan Sperber, David Premack, and Ann James Premack, 351–94. New York: Clarendon Press.

Hacking, Ian. 1999. *The Social Construction of What?* Cambridge, MA: Harvard University Press.

Ham, Yoo-Geun, Jeong-Hwan Kim, Jing-Jua Luo. 2019. "Deep Learning for Multi-Year ENSO Forecasts." *Nature* 573, 568–72.

Hamermesh, Daniel S. 2013. "Six Decades of Top Economics Publishing: Who and How?" *Journal of Economic Literature* 51, 162–72.

Han, E., M. Tan, E. Turk, D. Sridhar, G. Leung, K. Shibuya, N. Asgari, J. Oh, A. García-Basteiro, J. Hanefeld, A. Cook, L. Hsu, Y. Teo, D. Heymann, H. Clark, M. McKee, and H. Legido-Quigley. 2020. "Lessons Learnt From Easing COVID-19 Restrictions: an Analysis of Countries and Regions in Asia Pacific and Europe." *Lancet* 396, 1525–34.

Hannon, Michael. 2021. "Recent Work in the Epistemology of Understanding." *American Philosophical Quarterly* 58 (3), 269–90.

Hansen, Jens Ulrik, and Paula Quinon. 2023. "The Importance of Expert Knowledge in Big Data and Machine Learning." *Synthese* 201, 35.

Hardin, Garrett. 1968. "The Tragedy of the Commons." *Science* 162 (3859), 1243–8.

Hausman, Daniel. 1992. *The Inexact and Separate Science of Economics*. Cambridge, UK: Cambridge University Press.

Hausman, Daniel. 1997. "Why Does Evidence Matter So Little to Economic Theory?" In *Structures and Norms in Science*, edited by M. dalla Chiara, K. Doets, D. Mundici, and J. van Benthem, 395–407. Dordrecht: Kluwer.

Hausman, Daniel. 2007. "Why Look Under the Hood?" In *The Philosophy of Economics: An Anthology*, edited by Daniel Hausman, 183–7. Cambridge: Cambridge University Press.

Hedström, Peter, and Petri Ylikoski. 2010. "Causal Mechanisms in the Social Sciences." *Annual Review of Sociology* 36, 49–67.

Hempel, Carl. 1942. "The Function of General Laws in History." *Journal of Philosophy* 39 (2), 35–48.

Hempel, Carl, and Paul Oppenheim. 1948. "Studies in the Logic of Explanation." *Philosophy of Science* 15 (2), 135–75.

Henschen, Tobias. 2018. "The In-Principle Inconclusiveness of Causal Evidence in Macroeconomics." *European Journal for Philosophy of Science* 8 (3), 709–33.

Herfeld, Catherine. 2018. "Explaining Patterns, Not Details: Reevaluating Rational Choice Models in Light of Their Explananda." *Journal of Economic Methodology* 25 (2), 179–209.

Herzog, Lisa, and Bernardo Zacka. 2019. "Fieldwork in Political Theory: Five Arguments for an Ethnographic Sensibility." *British Journal of Political Science* 49 (2), 763–84.

Hey, Tony, Stewart Tansley, and Kristin Tolle (eds). 2009. *The Fourth Paradigm: Data-Intensive Scientific Discovery*. Redmond, WA: Microsoft Research.

Hills, Alison. 2016. "Understanding Why." *Nous* 50 (4), 661–88.

Hirshleifer, Jack. 1985. "The Expanding Domain of Economics." *American Economic Review* 83 (3), 53–68.

Hitchcock, Christopher. 1996. "The Role of Contrast in Causal and Explanatory Claims." *Synthese* 107, 395–419.

Hitchcock, Christopher, and James Woodward. 2003. "Explanatory Generalizations, Part II: Plumbing Explanatory Depth." *Nous* 37 (2), 181–99.

Hoover, Kevin. 2009. "Microfoundations and the Ontology of Macroeconomics." In *The Oxford Handbook of Philosophy of Economics*, edited by Don Ross and Harold Kincaid, 386–409. Oxford: Oxford University Press.

Horn, Christian F., Michael Ohneberg, Bjoern S. Ivens, and Alexander Brem. 2014. "Prediction Markets—a Literature Review Following Tziralis and Tatsiopoulos." *Journal of Prediction Markets* 8 (2), 89–126.

Howick, Jeremy, Paul Glasziou, and Jeffrey K. Aronson. 2013. "Problems with Using Mechanisms to Solve the Problem of Extrapolation." *Theoretical Medicine and Bioethics* 34, 275–91.

Humphreys, Paul. 2000. "Analytic versus Synthetic Understanding." In *Science, Explanation, and Rationality: The Philosophy of Carl G. Hempel*, edited by James H. Fetzer, 267–86. Oxford: Oxford University Press.

Huntington-Klein, Nick, Andreu Arenas, Emily Beam, Marco Bertoni, Jeffrey R. Bloem, Pralhad Burli, Naibin Chen, Paul Grieco, Godwin Ekpe, Todd Pugatch, Martin Saavedra, and Yaniv Stopnitzky. 2021. "The Influence of Hidden Researcher Decisions in Applied Microeconomics." *Economic Inquiry* 59, 944–60.

Hutcherson, Cendri A., Konstantyn Sharpinskyi, Michael E. W. Varnum, Amanda Rotella, Alexandra S. Wormley, Louis Tay, and Igor Grossmann. 2023. "On the Accuracy, Media Representation, and Public Perception of Psychological Scientists' Judgments of Societal Change." *American Psychologist*, February. DOI: 10.1037/amp0001151.

Hutchins, Edwin. 1995. *Cognition in the Wild*. Cambridge, MA: MIT Press.

Illari, Phyllis, and Jon Williamson. 2012. "What Is a Mechanism? Thinking about Mechanisms across the Sciences." *European Journal for Philosophy of Science* 2, 119–35.

Issenberg, Sasha. 2016. *The Victory Lab*. New York: Broadway.

Jacobs, David R., and Linda C. Tapsell. 2013. "Food Synergy: The Key to a Healthy Diet." *Proceedings of the Nutrition Society* 72, 200–6.

Japec, Lilli, Frauke Kreuter, Marcus Berg, Paul Biemer, Paul Decker, Cliff Lampe, Julia Lane, Cathy O'Neil, and Abe Usher. 2015. "Big Data in Survey Research: AAPOR Task Force Report." *Public Opinion Quarterly* 79 (4), 839–80.

Jarić, Ivan, Tina Heger, Federico Castro Monzon, Jonathan M. Jeschke, Ingo Kowarik, Kim R. McConkey, Petr Pyšek, Alban Sagouis, and Franz Essl. 2019. "Crypticity in biological invasions." *Trends in Ecology and Evolution* 34 (4), 291–302.

Jiménez-Buedo, Maria. 2021. "Reactivity in Social Scientific Experiments: What Is It and How Is It Different (and Worse) Than a Placebo Effect?" *European Journal for Philosophy of Science* 11, 42.

Jung, T., G. Balsamo, P. Bechtold, A. Beljaars, M. Koehler, M. Miller, J-J. Morcrette, A. Orr, M. Rodwell, and A. Tompkins. 2010. "The ECMWF Model Climate: Recent Progress Through Improved Physical Parametrizations." *Quarterly Journal of the Royal Meteorological Society* 136, 1145–60.

Kagel, John, and Alvin Roth (eds). 2016. *The Handbook of Experimental Economics, Volume 2*. Princeton: Princeton University Press.

Kaiserman, Alex. 2018. "'More of a Cause': Recent Work on Degrees of Causation and Responsibility." *Philosophy Compass* 13 (7), e12498.

Katz, Lawrence F., Jeffrey R. Kling, and Jeffrey B. Liebman. 2001. "Moving to Opportunity in Boston: Early Results of a Randomized Mobility Experiment." *Quarterly Journal of Economics* 116 (2), 607–54.

Kerwin, Jason. 2021. "Nothing Scales." Blog post, November 3, accessed October 26, 2022. Nothing Scales–Jason Kerwin

Khalidi, Muhammad A. 2010. "Interactive Kinds." *British Journal for the Philosophy of Science* 61, 335–60.

Khalidi, Muhammad A. 2013. *Natural Categories and Human Kinds: Classification in the Natural and Social Sciences*. Cambridge University Press.

Khalifa, Kareem. 2017. *Understanding, Explanation, and Scientific Knowledge*. New York: Cambridge University Press.

Khosrowi, Donal. 2019a. "Extrapolation of Causal Effects—Hopes, Assumptions, and the Extrapolator's Circle." *Journal of Economic Methodology* 26 (1), 45–58.

Khosrowi, Donal. 2019b. "Extrapolating Policy Effects." PhD dissertation, University of Durham.

Kincaid, Harold. 1996. *Philosophical Foundations of the Social Sciences*. Cambridge: Cambridge University Press.

Kitcher, Philip. 1989. "Explanatory Unification and the Causal Structure of the World." In *Scientific Explanation*, edited by Philip Kitcher and Wesley Salmon, 410–505. Minneapolis: University of Minnesota Press.

King, Gary, Robert Keohane, and Sidney Verba. 1994. *Designing Social Inquiry: Scientific Inference in Qualitative Research*. Princeton: Princeton University Press.

Kling, Jeffrey, Jeffrey Liebman, and Lawrence Katz. 2004. "Bullets Don't Got No Name: Consequences of Fear in the Ghetto." In *Discovering Successful Pathways in Children's Development: Mixed Methods in the Study of Childhood and Family Life*, edited by Thomas S. Weisner. Chicago: The University of Chicago Press.

Knüsel, Benedikt, Marius Zumwald, Christoph Baumberger, Gertrude Hirsch Hadorn, Erich M. Fischer, David N. Bresch, and Reto Knutti. 2019. "Applying Big Data Beyond Small Problems in Climate Research." *Nature Climate Change* 9, 196–202.

Kuhn, Thomas. 1962. *The Structure of Scientific Revolutions*. University of Chicago Press.

Kuipers, Theo (ed). 1987. *What Is Closer-to-the-Truth?* Amsterdam: Rodopi.

Kuorikoski, Jaakko, Aki Lehtinen, and Caterina Marchionni. 2010. "Economic Modelling as Robustness Analysis." *British Journal for the Philosophy of Science* 61, 541–67.

Kvanvig, Jonathan. 2003. *The Value of Knowledge and the Pursuit of Understanding*. Cambridge: Cambridge University Press.

Laimann, Jessica. 2020. "Capricious Kinds." *British Journal for the Philosophy of Science* 71, 1043–68.

Lakatos, Imre. 1970. "Falsification and the Methodology of Scientific Research Programmes." In *Criticism and the Growth of Knowledge*, edited by Imre Lakatos and Alan Musgrave, 91–196. Cambridge: Cambridge University Press.

Lander, Eric S. 2016. "The Heroes of CRISPR." *Cell* 164, 18–28.

Laudan, Larry. 1977. *Progress and Its Problems*. Berkeley and Los Angeles: University of California Press.

Lawler, Janet, and David Waldner. 2023. "Interpretivism versus Positivism in an Age of Causal Inference." In *Oxford Handbook of Philosophy of Political Science*, edited by Harold Kincaid and Jeroen van Bouwel, 221–42. Oxford University Press.

Lawson, Tony. 1997. *Economics and Reality*. London: Routledge.

Lazear, Edward P. 2000. "Economic Imperialism." *Quarterly Journal of Economics* 115 (1), 99–146.

Lazer, David, Ryan Kennedy, Gary King, and Alessandro Vespignani. 2014. "The Parable of Google Flu: Traps in Big Data Analysis." *Science* 343, 1203–5.

Leamer, Edward E. 2010. "Tantalus on the Road to Asymptotia." *Journal of Economic Perspectives* 24 (2), 31–46.

Lemoine, Philippe. 2021a. "The British Variant of SARS-CoV-2 and the Poverty of Epidemiology." CSPI blog, April 9. https://cspicenter.org/blog/waronscience/the-british-variant-of-sars-cov-2-and-the-povery-of-epidemiology/

Lemoine, Philippe. 2021b. "Is the Delta Variant Really More Than Twice as Transmissible as the Original Strain of the Virus?" CSPI blog, August 31. https://cspicenter.org/blog/waronscience/is-the-delta-variant-really-more-than-twice-as-transmissible-as-the-original-strain-of-the-virus/

Leonelli, Sabina. 2016. *Data-Centric Biology*. University of Chicago Press.

Levins, Richard. 1966. "The Strategy of Model Building in Population Biology." *American Scientist* 54 (4), 421–31.

Levitt, Steven D., and John A. List. 2007. "What Do Laboratory Experiments Measuring Social Preferences Reveal about the Real World?" *Journal of Economic Perspectives* 21 (2), 153–74.

Lewis, David. 1983. "New Work for a Theory of Universals." *Australasian Journal of Philosophy* 61 (4), 343–77.

Lewis-Kraus, Gideon. 2016. "The Great AI Awakening." *New York Times Magazine*, December 14.

Lipsitch, Marc. 2020. "We Know Enough Now to Act Decisively Against Covid-19. Social Distancing Is a Good Place to Start." *Stat News*, March 18.

Little, Daniel. 1991. *Varieties of Social Explanation*. Boulder: Westview.

Longino, Helen. 1990. *Science as Social Knowledge: Values and Objectivity in Scientific Inquiry*. Princeton: Princeton University Press.

Longino, Helen. 2002. *The Fate of Knowledge*. Princeton: Princeton University Press.

Loungani, Prakash. 2001. "How Accurate Are Private Sector Forecasts? Cross-Country Evidence from Consensus Forecasts of Output Growth." *International Journal of Forecasting* 17, 419–32.

Lucas, Robert. 2001. "Personal Memoir." unpublished manuscript.

Ludwig, David, and Stéphanie Ruphy. "Scientific Pluralism." In *The Stanford Encyclopedia of Philosophy (Winter 2021 Edition)*, edited by Edward N. Zalta. https://plato.stanford.edu/archives/win2021/entries/scientific-pluralism/.

Mackie, J. L. 1974. *The Cement of the Universe*. Oxford: Oxford University Press.

Magnus, P. D. 2012. *Scientific Enquiry and Natural Kinds: From Planets to Mallards*. Basingstoke: Palgrave MacMillan.

Magnus, P. D, and Craig Callender. 2004. "Realist Ennui and the Base Rate Fallacy." *Philosophy of Science* 71, 320–38.

Mäki, Uskali. 1992. "On the Method of Isolation in Economics." *Poznan Studies in the Philosophy of the Sciences and the Humanities* 26, 19–54.

Mallon, Ron. 2016. *The Construction of Human Kinds.* Oxford: Oxford University Press.

Mance, Henry, and David Sheppard. 2023. "The Science of Forecasting Ever More Extreme Weather." *Financial Times*, August 3. The science of forecasting ever more extreme weather | Financial Times (ft.com)

Manzo, Gianluca. 2020. "Complex Social Networks are Missing in the Dominant COVID-19 Epidemic Models." *Sociologica* 14 (1), 31–49.

Martin, Raymond. 1989. *The Past within Us: An Empirical Approach to Philosophy of History.* Princeton: Princeton University Press.

Maslen, Cei. 2004. "Causes, Contrasts and the Nontransitivity of Causation." In *Causation and Counterfactuals*, edited by John Collins, Ned Hall, and L. A. Paul, 341–57. Massachusetts: MIT Press.

Massimi, Michela. 2022. *Perspectival Realism.* Oxford: Oxford University Press.

May, Robert. 1976. "Simple Mathematical Models with Very Complicated Dynamics." *Nature* 261, 459–67.

May, Robert. 1977. "Thresholds and Breakpoints in Ecosystems with a Multiplicity of Stable States." *Nature* 269, 471–77.

Mayer-Schoenberger, Viktor, and Kenneth Cukier. 2013. *Big Data.* John Murray.

Meese, Richard A., and Kenneth Rogoff. 1983. "Empirical Exchange Rate Models of the Seventies: Do They Fit Out of Sample?" *Journal of International Economics* 14, 3–24.

Merton, Robert. 1967. *On Theoretical Sociology.* New York: Free Press.

Merton, Robert. 1968. *Social theory and social structure.* New York: Free Press.

Mill, J. S. 1843. *System of Logic.* London: Parker.

Miller, Boaz. 2016. "What Is Hacking's Argument for Entity Realism?" *Synthese* 193, 991–1006.

Miller, David. 1994. *Critical Rationalism: A Restatement and Defence.* Chicago: Open Court.

Mitchell, Sandra. 2000. "Dimensions of Scientific Law." *Philosophy of Science* 67, 242–65.

Möllenkamp, Meilin, Maike Zeppernick, and Jonas Schreyögg. 2019. "The Effectiveness of Nudges in Improving the Self-Management of Patients with Chronic Diseases: A Systematic Literature Review." *Health Policy* 123 (12), 1199–1209.

Montgomery, Kathryn. 2006. *How Doctors Think.* New York: Oxford University Press.

Morgan, Mary. 2002. "Model Experiments and Models in Experiments." In *Model-Based Reasoning: Science, Technology, Values*, edited by Lorenzo Magnani and Nancy J. Nersessian. New York: Kluwer Academic/Plenum Publishers.

Morgan, Mary. 2012. "Case Studies: One Observation or Many? Justification or Discovery?" *Philosophy of Science* 79 (5), 667–77.

Morgan, Mary, and Margaret Morrison (eds). 1999. *Models as Mediators: Perspectives on Natural and Social Science.* Cambridge: Cambridge University Press.

Mukherjee, Siddhartha. 2010. *The Emperor of All Maladies.* New York: Scribner.

Mullainathan, Sendhil, and Jann Spiess. 2017. "Machine Learning: An Applied Econometric Approach." *Journal of Economic Perspectives* 31 (2), 87–106.

Nagatsu, Michiro, and Judith Favereau. 2020. "Two Strands of Field Experiments in Economics: A Historical-Methodological Analysis." *Philosophy of the Social Sciences* 50 (1), 45–77.

National Academy of Sciences, Institute of Medicine. 2008. "Science, Evolution, and Creationism." Washington, DC: National Academies Press.

Neale, Elizabeth P., and Linda C. Tapsell. 2019. "Perspective: The Evidence-Based Framework in Nutrition and Dietetics: Implementation, Challenges, and Future Directions." *Advances in Nutrition* 10, 1–8.

Nelson, Alan. 1990. "Are Economics Kinds Natural?" In *Scientific Theories*, edited by C. Wade Savage, 102–135. Minneapolis: University of Minnesota Press.

Niiniluoto, Ilkka. 1987. *Truthlikeness*. Dordrecht: Reidel.

Nijhuis, Michelle. 2021. *Beloved Beasts: Fighting for Life in an Age of Extinction*. New York: Norton.

Northcott, Robert. 2008a. "Causation and Contrast Classes." *Philosophical Studies* 39 (1), 111–23.

Northcott, Robert. 2008b. "Weighted Explanations in History." *Philosophy of the Social Sciences* 38 (1), 76–96.

Northcott, Robert. 2012. "Partial Explanations in Social Science." In *Oxford Handbook of Philosophy of Social Science*, edited by Harold Kincaid, 130–53. Oxford University Press.

Northcott, Robert. 2013a. "Verisimilitude: A Causal Approach." *Synthese* 190 (9), 1471–88.

Northcott, Robert. 2013b. "Degree of Explanation." *Synthese* 190 (15), 3087–105.

Northcott, Robert. 2015. "Opinion Polling and Election Predictions." *Philosophy of Science* 82, 1260–71.

Northcott, Robert. 2017. "When Are Purely Predictive Models Best?" *Disputatio* 9 (47), 631–56.

Northcott, Robert. 2018. "The Efficiency Question in Economics." *Philosophy of Science* 85 (5), 1140–51.

Northcott, Robert. 2019. "Prediction versus Accommodation in Economics." *Journal of Economic Methodology* 26 (1), 59–69.

Northcott, Robert. 2020. "Big Data and Prediction: Four Case Studies." *Studies in the History and Philosophy of Science* 86, 96–104.

Northcott, Robert. 2021. "Pre-Emption Cases May Support, Not Undermine, the Counterfactual Theory of Causation." *Synthese* 198 (1), 537–55.

Northcott, Robert. 2022a. "Pandemic Modeling, Good and Bad." *Philosophy of Medicine* 3 (1), 1–20.

Northcott, Robert. 2022b. "Reflexivity and Fragility." *European Journal for Philosophy of Science* 12, 43.

Northcott, Robert. 2022c. "Economic Theory and Empirical Science." In *Routledge Handbook of Philosophy of Economics*, edited by Conrad Heilmann and Julian Reiss, 387–96. Routledge.

Northcott, Robert. 2023. "Prediction, History, and Political Science." In *Oxford Handbook of Philosophy of Political Science*, edited by Harold Kincaid and Jeroen van Bouwel, 452–66. Oxford University Press.

Northcott, Robert, and Anna Alexandrova. 2013. "It's Just a Feeling: Why Economic Models Do Not Explain." *Journal of Economic Methodology* 20, 262–67.

Northcott, Robert, and Anna Alexandrova. 2015. "Prisoner's Dilemma Doesn't Explain Much." In *The Prisoner's Dilemma*, edited by Martin Peterson, 64–84. Cambridge: Cambridge University Press.

Norton, John. 2021. *The Material Theory of Induction*. University of Calgary Press. https://doi.org/10.2307/j.ctv25wxcb5

Nowak, Martin A., and Karl Sigmund. 1999. "Phage-Lift for Game Theory." *Nature* 398, 367–8.

Oddie, Graham. 1986. *Likeness to Truth*. Dordrecht: Reidel.

Okruhlik, Kathleen. 1994. "Gender and the Biological Sciences." *Canadian Journal of Philosophy* 20, 21–42.

Pawson, Ray, Trisha Greenhalgh, Gill Harvey, and Kieran Walshe. 2005. "Realist Review: A New Method of Systematic Review Designed for Complex Policy Interventions." *Journal of Health Services Research and Policy* 10 (1), 21–34.

Pawson, Ray, and Nicholas Tilley (1997). "An Introduction to Scientific Realist Evaluation." In *Evaluation for the 21st Century: A Handbook*, edited by Eleanor Chelimsky and William R. Shadish, 405–18. Washington: Sage Publications, Inc.

Pearl, Judea. 2009. *Causality*. Cambridge: Cambridge University Press.

Pemberton, John, and Nancy Cartwright. 2014. "Ceteris Paribus Laws Need Machines to Generate Them." *Erkenntnis* 79 (10), 1745–58.

Pietsch, Wolfgang. 2015. "Aspects of Theory-Ladenness in Data-Intensive Science." *Philosophy of Science* 82, 905–16.

Pietsch, Wolfgang. 2016. "The Causal Nature of Modeling with Big Data." *Philosophy and Technology* 29, 137–71.

Pietsch, Wolfgang. 2021. *Big Data*. Elements in the Philosophy of Science. Cambridge: Cambridge University Press.

Plott, Charles R. 1997. "Laboratory Experimental Testbeds: Application to the PCS Auction." *Journal of Economics and Management Strategy* 6 (3), 605–38.

Plutynski, Anya. 2018. *Explaining Cancer: Finding Order in Disorder*. New York: Oxford University Press.

Pollack, Simon. 2022. "Epistemic Risk and the Influence of Values in Actuarial Modelling." MA dissertation, Birkbeck, University of London.

Popper, Karl. 1972. *Objective Knowledge: An Evolutionary Approach*. Oxford: Clarendon Press.

Popper, Karl. 1989. *Conjecture and Refutations*. London: Routledge.

della Porta, Donatella. 1995. *Social Movements, Political Violence, and the State*. Cambridge University Press.

Potochnik, Angela. 2017. *Idealization and the Aims of Science*. Chicago: University of Chicago Press.

Price, Justin. 2019. "The Landing Zone: Ground for Model Transfer in Chemistry." *Studies in the History and Philosophy of Science* 77, 21–8.

Ravuri, S., K. Lenc, M. Willson, D. Kangin, R. Lam, P. Mirowski, M. Fitzsimons, M. Athanassiadou, S. Kashem, S. Madge, R. Prudden, A. Mandhane, A. Clark, A. Brock, K. Simonyan, R. Hadsell, N. Robinson, E. Clancy, A. Arribas, and S. Mohamed. 2021. "Skilful Precipitation Nowcasting Using Deep Generative Models of Radar." *Nature* 597, 672–77.

Reiss, Julian. 2008. *Error in Economics: Towards a More Evidence-Based Methodology*. London: Routledge.

Reiss, Julian. 2012. "Idealization and the Aims of Economics: Three Cheers for Instrumentalism." *Economics and Philosophy* 28, 363–83.

Reiss, Julian. 2013a. "Contextualising Causation: Part I." *Philosophy Compass* 8 (11), 1066–75.

Reiss, Julian. 2013b. "Contextualising Causation: Part II." *Philosophy Compass* 8 (11), 1076–90.

Reiss, Julian. 2017. "Are There Social Scientific Laws?" In *The Routledge Companion to Philosophy of Social Science,* edited by Lee McIntyre and Alexander Rosenberg, 295–309. New York: Routledge.

Reiss, Julian. 2019. "Expertise, Agreement, and the Nature of Social Scientific Facts or: Against Epistocracy." *Social Epistemology* 33 (2), 183–92.

Rejmanek, Marcel, and David Richardson. 1996. "What Attributes Make Some Plant Species More Invasive?" *Advances in Invasion Ecology* 77 (6), 1655–61.

Richardson, David, and Marcel Rejmanek. 2004. "Conifers as Invasive Aliens: A Global Survey and Predictive Framework." *Diversity and Distributions* 10, 321–31.

Robbins, Lionel. 1935. *An Essay on the Nature and Significance of Economic Science.* London: MacMillan.

Rodrik, Dani. 2015. *Economics Rules: The Rights and Wrongs of the Dismal Science.* New York: Norton.

Rosenberg, Alexander. 1992. *Economics: Mathematical Politics or Science of Diminishing Returns?* Chicago: University of Chicago Press.

Ross, Don. 2014. *Philosophy of Economics.* Palgrave MacMillan.

Roth, Alvin E. 2002. "The Economist as Engineer: Game Theory, Experimentation, and Computation as Tools for Design Economics." *Econometrica* 70 (4), 1341–78.

The Royal Society DELVE Initiative. 2020. 'SARS-CoV-2 Vaccine Development & Implementation; Scenarios, Options, Key Decisions', DELVE Report No. 6, October 1.

Rubin, Ashley. 2021. *Rocking Qualitative Social Science: An Irreverent Guide to Rigorous Research.* Stanford: Stanford University Press.

Ruiz, Nadia, and Armin W. Schulz. 2023. "Microfoundations and Methodology: a Complexity-Based Reconceptualization of the Debate." *British Journal for the Philosophy of Science* 74 (2), 359–79.

Runhardt, Rosa. 2015. "Evidence for Causal Mechanisms in Social Science: Recommendations from Woodward's Manipulability Theory of Causation." *Philosophy of Science* 85(5), 1296–1307.

Rutter, H., N. Savona, K. Glonti, J. Bibby, S. Cummins, D. Finegood, F. Greaves, L. Harper, P. Hawe, L. Moore, M. Petticrew, E. Rehfuess, A. Shiell, J. Thomas, and M. White. 2017. "The Need for a Complex Systems Model of Evidence for Public Health." *The Lancet* 390 (10112), 2602–4.

Saatsi, Juha. 2017. "Replacing Recipe Realism." *Synthese* 194, 3233–44.

Sagoff, Mark. 2016. "Are There General Causal Forces in Ecology?" *Synthese* 193, 3003–24.

Salmon, Wesley. 1981. "Rational Prediction." *British Journal for the Philosophy of Science* 32 (2), 115–25.

Schaffer, Jonathan. 2005. "Contrastive Causation." *Philosophical Review* 114 (3), 297–328.

Schaffer, Jonathan. 2012. "Causal Contextualisms." In *Contrastivism in Philosophy*, edited by Martijn Blaauw, 35–63. London: Routledge.

Schön, Donald. 1992. *The Reflective Practitioner: How Professionals Think in Action.* London: Routledge.

Schroeder, S. Andrew. 2021. "How to Interpret COVID-19 Predictions: Reassessing the IHME's Model." *Philosophy of Medicine* 2 (1), 1–7.

Scott, James C. 1999. *Seeing Like a State: How Certain Schemes to Improve the Human Condition Have Failed.* Yale University Press.

Scriven, Michael. 1956. "A Possible Distinction Between Traditional Scientific Disciplines and the Study of Human Behavior." In *The Foundations of Science and the Concepts of Psychology and Psychoanalysis*, edited by Herbert Feigl and Michael Scriven, 330–9. Minneapolis: University of Minnesota Press.

Seckinelgin, Hakan. 2017. *The Politics of Global AIDS: Institutionalization of Solidarity, Exclusion of Context.* Switzerland: Springer International Publishing.

Shan, Yafeng. 2019. "A New Functional Approach to Scientific Progress." *Philosophy of Science* 86, 739–58.

Silver, Nate. 2012. *The Signal and the Noise.* Penguin.

Silver, Nate. 2014. "How the FiveThirtyEight Senate Forecast Model Works." September 17. http://fivethirtyeight.com/features/how-the-fivethirtyeight-senate-forecast-model-works/

Skyrms, Brian. 2014. *Evolution of the Social Contract*. Cambridge: Cambridge University Press.

Smith, George. 2014. "Closing the Loop: Testing Newtonian Gravity, Then and Now." In *Newton and Empiricism*, edited by Zvi Biener and Eric Schliesser, 262–352. New York: Oxford University Press.

Smith, Noah. 2022. "Macroeconomics Is Still in Its Infancy." Noahpinion substack, November 8. Macroeconomics is still in its infancy—by Noah Smith (substack.com)

Sober, Elliott. 1999. "Testability." *Proceedings and Addresses of the American Philosophical Association* 73, 47–76.

Sober, Elliott. 2009. "Absence of Evidence and Evidence of Absence: Evidential Transitivity in Connection with Fossils, Fishing, Fine-Tuning, and Firing Squads." *Philosophical Studies* 143, 63–90.

Solomon, Miriam. 2015. *Making Medical Knowledge*. New York: Oxford University Press.

SPI-M-O (Scientific Pandemic Influenza Group on Modelling, Operational Sub-group). 2020. "Consensus view on Covid-19." March 20. https://assets.publishing.service.gov.uk/government/uploads/system/uploads/attachment_data/file/887463/24-spi-m-o-consensus-view-20032020.pdf, accessed May 12, 2022.

Spirtes, Peter, Clark Glymour, and Richard Scheines. 2000. *Causation, Prediction, and Search*. Cambridge, MA: MIT Press.

Steel, Daniel. 2008. *Across the Boundaries: Extrapolation in Biology and Social Science*. New York: Oxford University Press.

Stegenga, Jacob. 2018. *Medical Nihilism*. New York: Oxford University Press.

Steinberger, Florian. 2019. "Three Ways in Which Logic Might Be Normative." *Journal of Philosophy* 116 (1), 5–31.

Sterelny, Kim. 2016. "Contingency and History." *Philosophy of Science* 83 (4), 521–39.

Strevens, Michael. 2012. "Ceteris Paribus Hedges: Causal Voodoo That Works." *Journal of Philosophy* 109, 652–75.

Strogatz, Steven. 2020. *Infinite Powers: The Story of Calculus, the Language of the Universe*. New York: Houghton Mifflin.

Sturgis, Patrick, Nick Baker, Mario Callegaro, Stephen Fisher, Jane Green, Will Jennings, Jouni Kuha, Ben Lauderdale, and Patten Smith. 2016. *Report of the Inquiry into the 2015 British General Election Opinion Polls*. London: Market Research Society and British Polling Council.

Sugihara, George, Robert May, Hao Ye, Chih-Hao Hsieh, Ethan Deyle, Michael Fogarty, and Stephan Munch. 2012. "Detecting Causality in Complex Ecosystems." *Science* 338 (6106), 496–500.

Sunstein, Cass R. 2007. "Of Montreal and Kyoto: A Tale of Two Protocols." *Harvard Environmental Law Review* 31 (1).

Syrjänen, Pekka. 2022. "The Epistemic Role of Prediction in Science." PhD dissertation, University of Helsinki.

Taylor, Charles. 1971. "Interpretation and the Sciences of Man." *Review of Metaphysics* 25, 3–51.

Teller, Paul. 2001. "Twilight of the Perfect Model Model." *Erkenntnis* 55 (3), 393–415.

Tetlock, Philip. 2005. *Expert Political Judgment: How Good Is It? How Can We Know?* Princeton: Princeton University Press.

Tetlock, Philip, and Dan Gardner. 2015. *Superforecasting: The Art and Science of Prediction*. New York: Crown Publishing Group.

Thucydides (1974/c 400 BCE). *The History of the Peloponnesian War*. Translated by Rex Warner. London: Penguin.

Tolstoy, Leo. 1967/1862. "On Teaching the Rudiments." In *Tolstoy on Education*, edited and translated by Leo Wiener. Chicago and London: University of Chicago Press.

Tucker, Aviezer. 2004. *Our Knowledge of the Past*. Cambridge: Cambridge University Press.

Turner, Derek, and Michelle Turner. 2021. ""I'm Not Saying It Was Aliens": An Archeological and Philosophical Analysis of a Conspiracy Theory." In *Explorations in Archeology and Philosophy*, edited by Anton Killin and Sean Allen-Hermanson, 7–24. Springer Verlag.

Tziralis, Georgios, and Ilias P. Tatsiopoulos. 2007. "Prediction Markets: An Extended Literature Review." *Journal of Prediction Markets* 1, 75–91.

van Basshuysen, Philippe, and Lucie White. 2021a. "Were Lockdowns Justified? A Return to the Facts and Evidence." *Kennedy Institute of Ethics Journal* 31 (4), 405–28.

van Basshuysen, Philippe, and Lucie White. 2021b. "The Epistemic Duties of Philosophers: an Addendum." *Kennedy Institute of Ethics Journal* 31 (4), 447–51.

van den Assem, Martijn J., Dennie van Dolder, and Richard H. Thaler. 2012. "Split or Steal: Cooperative Behavior When the Stakes Are Large." *Management Science* 58 (1), 2–20.

van Fraassen, Bas. 1980. *The Scientific Image*. Oxford: Oxford University Press.

van Fraassen, Bas. 1989. *Laws and Symmetry*. Oxford: Oxford University Press.

Vickers, Peter. 2022. *Identifying Future-Proof Science*. Oxford: Oxford University Press.

Vivalt, Eva. 2020. "How Much Can We Generalize from Impact Evaluations?" *Journal of the European Economic Association* 18 (6), 3045–89.

Volz, E., S. Mishra, M. Chand, J. Barrett, R. Johnson, L. Geidelberg, W. Hinsley, D. Laydon, G. Dabrera, Á. O'Toole, R. Amato, M. Ragonnet-Cronin, I. Harrison, B. Jackson, C. Ariani, O. Boyd, N. Loman, J. McCrone, S. Gonçalves, D. Jorgensen, R. Myers, V. Hill, D. Jackson, K. Gaythorpe, N. Groves, J. Sillitoe, D. Kwiatkowski, The COVID-19 Genomics UK (COG-UK) consortium, S. Flaxman, O. Ratmann, S. Bhatt, S. Hopkins, A. Gandy, A. Rambaut, and N. Ferguson. 2020. "Assessing Transmissibility of SARS-CoV-2 Lineage B.1.1.7 in England." *Nature* 593, 266–9.

Ward, Michael D., Brian D. Greenhill, and Kristin M. Bakke. 2010. "The Perils of Policy by P-value: Predicting Civil Conflicts." *Journal of Peace Research* 47 (4), 363–75.

Watkins, John. 1968. "Non-inductive corroboration." In *The Problem of Inductive Logic*, edited by Imre Lakatos, 61–6. Amsterdam: North-Holland Publishing.

Weber, Marcel. "Experiment in Biology." In *The Stanford Encyclopedia of Philosophy* (Summer 2018 Edition), edited by Edward N. Zalta. https://plato.stanford.edu/archives/sum2018/entries/biology-experiment/.

Weisberg, Michael. 2013. *Simulation and Similarity: Using Models to Understand the World*. Oxford: Oxford University Press.

Wells, Anthony. 2018. "Why the Polls Were Wrong in 2017." Blog post, June 7. Why the polls were wrong in 2017 | UK Polling Report

Westerblad, Oscar. 2023. "Making Sense of Understanding: A Pragmatist Account of Scientific Understanding." PhD dissertation, University of Cambridge.

White, Lucie, Philippe van Basshuysen, and Mathias Frisch. 2022. "When Is Lockdown Justified?" *Philosophy of Medicine* 3 (1), 1–22. https://doi.org/10.5195/pom.2022.85.

Wilson, Robert. 2004. *Genes and the Agents of Life: The Individual in the Fragile Sciences—Biology*. New York: Cambridge University Press.

Wilson, Robert. 2005. *Boundaries of the Mind: The Individual in the Fragile Sciences—Cognition*. New York: Cambridge University Press.

Winch, Peter. 1958. *The Idea of a Social Science*. Routledge.

Winsberg, Eric, Jason Brennan, and Chris W. Surprenant. 2020. "How Government Leaders Violated Their Epistemic Duties during the SARS-CoV-2 Crisis." *Kennedy Institute of Ethics Journal* 30 (3–4), 215–42.

Winsberg, Eric, Jason Brennan, and Chris W. Surprenant. 2021. "This Paper Attacks a Strawman but the Strawman Wins: A reply to van Basshuysen and White." *Kennedy Institute of Ethics Journal* 31 (4), 429–46.

Woodward, James. 2003. *Making Things Happen: A Theory of Causal Explanation.* New York: Oxford University Press.

Woodward, James. 2006. "Sensitive and Insensitive Causation." *Philosophical Review* 115, 1–50.

Ylikoski, Petri. 2019. "Mechanism-Based Theorizing and Generalization from Case Studies." *Studies in the History and Philosophy of Science* 78, 14–22.

Ylikoski, Petri. 2021. "Understanding the Coleman Boat." In *Research Handbook on Analytical Sociology,* edited by Gianluca Manzo, 49–63. Cheltenham: Edward Elgar.

Ylikoski, Petri, and Emrah Aydinonat. 2014. "Understanding with theoretical models." *Journal of Economic Methodology* 21, 19–36.

Ylikoski, Petri, and Jaakko Kuorikoski. 2010. "Dissecting Explanatory Power." *Philosophical Studies* 148, 201–19.

Ylikoski, Petri, and Julie Zahle. 2019. "Case Study Research in the Social Sciences." *Studies in the History and Philosophy of Science* 78, 1–4.

Youson, Matt. 2020. "The Insider's Guide to . . . F1 Car Development." February 10, www. formula1.com. The insider's guide to . . . F1 car development | Formula 1®

Index

For the benefit of digital users, indexed terms that span two pages (e.g., 52–53) may, on occasion, appear on only one of those pages.

Tables are indicated by an italic *t* following the page number.